INDEX ON CENSORSHIP 1 1996

Volume 25 No 1 January/February 1996   Issue 168

Editor & Chief Executive
Ursula Owen
Director of Administration
Philip Spender
Deputy Editor
Judith Vidal-Hall
Production Editor
Rose Bell
Fundraising Manager
Elizabeth Twining
News Editor
Adam Newey
Editorial Assistants
Anna Feldman
Philippa Nugent
Africa
Adewale Maja-Pearce
Eastern Europe
Irena Maryniak
Circulation & Marketing Director
Louise Tyson
Subscriptions Manager
Kelly Cornwall

Accountant
Suzanne Doyle
Volunteer Assistants
Michaela Becker
Thomasin Brooker
Laura Bruni
Nathalie de Broglio
Erum Faruqi
Ian Franklin
Joe Hipgrave
Dionne King
Anne Logie
Nicholas McAulay
Jamie McLeish
Albert Opoku
Grazia Pelosi
Sarah Smith
James Solomon
Katheryn Thal
Saul Venit
Henry Vivian-Neal
Tara Warren
Predrag Zivkovic

Directors
Louis Blom-Cooper, Ajay Chowdhury, Caroline Moorehead, Ursula Owen, Peter Palumbo, Jim Rose, Anthony Smith, Philip Spender, Sue Woodford (Chair)

Council
Ronald Dworkin, Amanda Foreman, Thomas Hammarberg, Clive Hollick, Geoffrey Hosking, Michael Ignatieff, Mark Littman, Pavel Litvinov, Robert McCrum, Uta Ruge, William Shawcross, Suriya Wickremasinghe

Patrons
Chinua Achebe, David Astor, Robert L Bernstein, Harold Evans, Richard Hamilton, Stuart Hampshire, Yehudi Menuhin, Iris Murdoch, Philip Roth, Tom Stoppard, Michael Tippett, Morris West

Australian committee
Philip Adams, Blanche d'Alpuget, Bruce Dawe, Adele Horin, Angelo Loukakis, Ken Methold, Laurie Muller, Robert Pullan and David Williamson c/o Ken Methold, PO Box 6073, Toowoomba West, Queensland 4350

Danish committee
Paul Grosen, Niels Barfoed, Claus Sønderkøge, Herbert Pundik, Nils Thostrup, Toni Liversage and Björn Elmquist, c/o Claus Sønderkøge, Utkaervej 7, Ejerslev, DK-7900 Nykobing Mors

Dutch committee
Maarten Asscher, Gerlien van Dalen, Christel Jansen, Bert Janssens, Chris Keulemans, Frank Ligtvoet, Hans Rutten, Mineke Schipper, Steffie Stokvis, Martine Stroo and Steven de Winter, c/o Gerlien van Dalen and Chris Keulemans, De Balie, Kleine-Gartmanplantsoen 10, 1017 RR Amsterdam

Norwegian committee
Trond Andreassen, Jahn Otto Johansen, Alf Skjeseth and Sigmund Strømme, c/o NFF, Skippergt. 23, 0154 Oslo

Swedish committee
Ana L Valdés and Gunilla Abrandt, c/o *Dagens Nyheter*, Kulturredaktionen, S-105 15 Stockholm, Sweden

USA committee
Susan Kenny, Peter Jennings, Jeri Laber, Anne Nelson, Harvey J Kaye, Wendy Wolf, Rea Hederman, Jane Kramer, Gara LaMarche, Faith Sale, Michael Scammell

Cover design by Andrea Purdie

Index on Censorship (ISSN 0306-4220) is published bimonthly by a non-profit-making company:
Writers & Scholars International Ltd,
Lancaster House,
33 Islington High Street, London N1 9LH
Tel: 0171-278 2313 Fax: 0171-278 1878
E-mail: indexoncenso@gn.apc.org
http://www.oneworld.org/index_oc/

Index on Censorship is associated with Writers & Scholars Educational Trust, registered charity number 325003

Second class postage (US subscribers only) paid at Irvington, New Jersey. Postmaster: send US address changes to *Index on Censorship* c/o Virgin Mailing & Distribution, 10 Camptown Road, Irvington, NJ 07111
Subscriptions 1996 (6 issues p.a.)
Individuals: UK £36, US $50, rest of world £42
Institutions: UK £40, US $70, rest of world £46
Students: UK £25, US $35, rest of world £31

© This selection Writers & Scholars International Ltd, London 1996
© Contributors to this issue, except where otherwise indicated

Printed by Martins, Berwick upon Tweed, UK

COVER PHOTOS: Front cover: *Department of Child Radiation Risk, Kyiv, Ukraine, 1994: child victim of Chernobyl:/Gareth Wyn Jones/Panos;* (back and title page) *Ken Saro-Wiwa/Norbert Michalke/Octopus*

Former Editors: Michael Scammell (1972-81); Hugh Lunghi (1981-83); George Theiner (1983-88); Sally Laird (1988-89); Andrew Graham-Yooll (1989-93)

EDITORIAL

When knowledge is not enough

IT'S BEEN a busy two months since we last went to press. On 10 November, the writer Ken Saro-Wiwa and his eight fellow Ogoni activists were executed in Nigeria (p57). The world reacted with horror and surprise, but for human rights workers it was no surprise. They knew he had been in prison for many months facing the death penalty, and had known for several years the volatility of the Ogoniland conflict that sent him to the gallows. But knowledge was not enough. Nor, in this case was publicity: the huge worldwide publicity came late, and much too late to save the nine. *Index* looks at the crisis in the human rights world, and at why human rights organisations, more professional and more prolific than ever before, are apparently failing to get their message across.

Yitzhak Rabin's killing on 4 November forces us to ask other questions — about extremism and hate speech, and their consequences. Some extremely violent anti-Rabin sentiments were granted a platform in the Israeli media (which generally supported the peace process). This was probably as much a ratings as a free speech issue. But hate speech in this case was transformed into bullets. Where does that leave the issue of free expression? These are difficult questions.

What is not difficult to see is that in the case of human rights versus *realpolitik*, the latter usually wins. The Commonwealth Conference refused to condemn Nigeria until it was too late. As for Israel, what was absent from the mourning for Rabin was the fact, as Stan Cohen says (p22), that the Oslo agreement itself is manifestly unjust to the Palestinians and that 'the choice is not between reversing or continuing Rabin's direction: the choice is whether to convert the easy kitsch of peace into the hard concessions of justice.'

Speaking of which, however desirable the end of the war in former Yugoslavia and however difficult the solutions, the fact is that in Dayton, Ohio, territorial imperatives once more elbowed human rights to the margins (p9). All of which makes our point — human rights are losing out, it seems increasingly so, and the human rights world needs to ask why, and what it can do about it. ❏

CONTENTS

EDITORIAL	3	**URSULA OWEN** When knowledge is not enough
APPRECIATION	6	**TED HUGHES** Stephen Spender: poet of a lost culture
DIARY	9	**W L WEBB** Bosnia: who goes home?
	17	Kosovo: unwanted occupation...
PHOTO FEATURE	18	**MELANIE FRIEND** ...unwilling refugees
LETTER FROM JERUSALEM	22	**STANLEY COHEN** The kitsch of death and peace
IN THE NEWS	28	**IRENA MARYNIAK** Russia: notes from prison
	30	**ISABEL HILTON** Tibet: Tibetan succession
	32	**KARIM ALRAWI** Egypt: military sidestep
HUMAN RIGHTS	35	**COMMUNICATING HUMAN RIGHTS**
	36	**STANLEY COHEN** Witnessing the truth
	46	**PADDY COULTER** Think pictures
	48	**ROSALINDA GALANG** Keeping rights relevant
	52	**CAROLINE MOOREHEAD & URSULA OWEN** Time to think again
LETTER FROM LAGOS	57	**ADEWALE MAJA-PEARCE** The murderers in our midst
USA DIVIDES	62	**VLADIMIR VOINOVICH** Gang of four
	71	**MUMIA ABU-JAMAL** The power of money
	73	**JUNE JORDAN** Big country, no nation
NETSCAPE	78	**BRIAN WINSTON & PAUL WALTON** Virtually free
	80	**KIRSTY GORDON** Companions of the superhighway
COVER STORY	85	**CHERNOBYL: ONCE AND FUTURE SHOCK**
	86	**ANTHONY TUCKER** Confusion and deceit
	94	**MIKHAIL BYCKAU** A liquidator's story
	96	**DAVID HEARST** Taming the beast
	101	**ÜMIT ÖZTÜRK** A little radiation does you good

CONTENTS

BABEL

105 **CHILDREN OF CHERNOBYL**

COUNTRY FILE

111 **BELARUS: NATION IN SEARCH OF A HISTORY**

112 **VERA RICH** 1991 and all that
119 **VERA RICH** Frantsisk Skaryna
120 **YUYA HURTAVENKO & T I ULEVICH** Saints and heroes
121 **ANDREW WILSON** In search of a history
126 **KASYA KAMOTSKAYA** President, go home!
127 **ANNA FELDMAN** Minsk, Pinsk and other places
128 **ALEXANDER BELY** Freedom of oppression
131 **ALAKSANDR LUKASHENKA** Press edict
132 **ZHANA LITVINA** No prospects
134 **ADAM GLOBUS** Demonicameron tales
140 **BOLHA IPATAVA** Crossing the line
142 **VALENTIN TARAS** The tale of Tuteishin

NATIVE AMERICA

145 **GORDON BROTHERSTON** Native testimony in the Americas
152 **HUMBERTO AK'ABAL** Poems

FICTION

153 **GUILLERMO MARTÍNEZ** Vast hell

LEGAL

162 **JULIAN PETLEY** Savoy scrapbook

INDEX INDEX

167 **FLASHPOINTS** Australia, China, Indonesia, USA

Index and **WSET** depend on donations
to guarantee their independence and to fund research

The Trustees and Directors would like to thank
all those whose donations support *Index on Censorship* and
Writers and Scholars Educational Trust, including

The Bromley Trust
Charity Know How
The Combined Charities Trust
The European Commission
The Norwegian Association of Literary Translators

APPRECIATION: STEPHEN SPENDER

TED HUGHES

Poet of lost culture

MY FRIENDSHIP with Stephen was, I am sure, marginal for him. Yet in several ways it was important for me. When he materialised to my late teens in the quartet, Auden, Spender, MacNeice, Day Lewis, he was the one of the four that attracted my curiosity most. Not because he seemed the most open and approachable, though he did, but because of something in his verse.

Later on, a more definite relationship was thrust upon us. When my first collection of verse won an Anglo-American prize, he was one of the judges. From this I assumed, naturally enough, that he took a benevolent interest in me. I say naturally, yet I made no such assumption about the other judges, Auden and Marianne Moore, though I had met Marianne Moore, and felt to get on with her, before I met Stephen, and I first met Auden where I first met Stephen, at a Faber party in 1960.

Perhaps I remained guarded at those first meetings, but Stephen saw what was needed and took the initiative. Somehow he instantly reassured me that my assumption about him was correct. Our friendship flourished from there. The fact that we rarely met and hardly ever corresponded seemed to make no difference. For the next 35 years he made me feel our friendship was special and growing constantly stronger. A friendship waiting, I felt, only for some shift of circumstances to make it a very close one. Circumstances never shifted in the right way. So after his death I was left with this sense of a special, long-lasting relationship that actually had very few concrete memories.

Maybe many others feel something similar. His famous charm could easily make you feel special. Most people responded to his vivacity, which almost glowed, like an aura of incandescence. He drew you immediately into his sphere, where you were enveloped in overflowing, generous, happy feelings, passionate sympathies and great sweetness.

APPRECIATION: STEPHEN SPENDER

But the complications could be heard in his voice. It came straight from the hot springs, a seething conflict of contraries: gleeful and serious, mischievous and anxiously concerned, globally burdened and gossipy, boyish and sage, a controlled, painstakingly probing surgery to get at the heart of things, counterpointed with an irrepressible Dionysiac need for irreverent laughter and jokes. Apart from the flow of his personal warmth, two things in our friendship meant a lot to me. One had to do with the culture to which he belonged. The other with a particular quality of his poetic sensibility — which to some degree pervaded all his writings, and was directly connected, I feel, to what I found most appealing in his personality.

He belonged to the broad European culture that fell in the First World War and received its *coup de grâce* in the Second. For his generation, the international cultural solidarities were still within one family. With the next generation of English poets, born around 1920, one sees how the pattern has disintegrated and their cultural frontier narrowed. For my generation, 10 years later, little of that old culture seemed to remain, except as it could be picked up in junk-shops. Or except as it survived in some of Stephen's generation. And, for me, in Stephen. That became one of his roles, in his later decades: to be the sole living English poet representative of that lost culture. For myself, I was always keenly aware of just how much that inheritance enriched and broadened his judgements. Knowing Stephen, and living in his affectionate regard, I felt to have some access to it, in a very real way. Perhaps what I'm saying is that his presence re-enforced one's nostalgia for that bigger vision. He recognised and gave confidence to those bits and pieces of it that one had managed to appropriate, looting the ruins. So he nurtured any effort to salvage and find a future for whatever still seemed valuable. In my exchanges with him, this particular wavelength always became highly activated — though it wouldn't be very easy to say precisely what was transmitted and exchanged. After his death, part of that great gap, for me, was the sense of the sudden loss of this lifeline to that older culture.

The particular quality of his poetic gift that I cherished first showed itself in what most readers of his early poems notice — a certain kind of sharp image, usually described as 'cinematic'. I saw it rather as a jagged splinter in a cut — in other words I felt a biological actuality in it, shocking, intimate, an open injury of some kind, above all painful.

This jagged splinter in its cut was evidence, I thought, of something more than hurt. Perhaps other elements in his poems prompted me to see it as evidence of a virtue, a spiritual strength. A vulnerability — but one that refused to protect itself. An innocence that had the courage in spite of everything to stay innocent. It seemed to me important and suggestive of possibilities for a poetry that Stephen himself neglected to develop — in his preoccupation with other directions. In an essay about Keith Douglas I remarked that Douglas had absorbed a good deal of Auden's language. I have always regretted failing to mention what I think he might have found even more useful. Stephen's exposed inner vulnerability, and especially the pain, the visual effect and lighting of that Spenderesque image, are exactly what gave Douglas direct access to what he most needed — a cinematic or as-if-seen-through-a-lens focus on the naked predicament of the subjective life — the psychic autobiography. One could argue that this mode of operation has had as much influence on subsequent English poetry as the more identifiable one of Auden's, if not more.

That kind of image became less characteristic of the styles Stephen moved towards, but the sensibility that had produced it went on unchanged beneath everything. And that sensibility was always palpably there, in his physical presence. In his conversation the controlled, painstakingly probing surgery that I mentioned was fundamental. He seemed to be questioning everything he said with his own inmost feeling — his own vulnerability, his own wounds, perhaps. The process was registered in every changing detail of his facial expression. And in every mercurial step of his voice — with its unforgettable range of tone, melody and inflection, painfully restrained by concern for the way things are, or engulfed by his sudden laugh.

Towards the end of his life I was glad to see him turning towards a more open, relaxed, autobiographical kind of poem. His psychic autobiography was the power-line beneath his worldly one, and these reminiscences begin to release it in the most lucid and fascinating way. I wish he had started on them earlier and written more. They impart something of the living secret of that incandescence which, when one met him, made one feel simultaneously so grateful, amused, refreshed and favoured. ❑

Ted Hughes' Difficulties of a Bride Groom *(short stories) and* Collected Animal Poems *volumes 1-4 were published by Faber in September 1995. His* Choice of Coleridge's Verse *will be published in March*

BOSNIAN DIARY

W L WEBB

Who goes home?

They made a peace in Dayton, Ohio. But in the Krajina villages still burn, the cities of Bosnia are flattened, the population is scattered on the winds of winter. To what and where do the refugees return?

SOMEWHERE in *Black Lamb and Grey Falcon*, her epic panorama of Balkan landscape, myth and prejudice, Rebecca West says that she had come to Yugoslavia 'to see what history meant in flesh and blood'. One flinches, reading that 60 years later, following in her footsteps in what must seem to the South Slavs just another such 'low, dishonest decade' as the '30s.

For the moment, their flesh and blood is mostly still safe in its skin, and not, thank God and His improbable agents lately in Dayton, Ohio, splashed around a market-place in Sarajevo, or the little square in Tuzla where boys and girls used to crowd of an evening to drink coffee and hope they were falling in love. Bosnia is still the sovereign place to go for such history lessons. It's also the place in which to contemplate the uses and abuses of cartography. What went on chiefly in the Bob Hope Hotel in Dayton, Ohio, was the endless drawing and tearing up of maps (even the final documents included an annexe of 102 of them), culminating in a session in 'the Nintendo room', the base's map-room, where fancy electronics reproduced the detail of obsessively fought-over territories down to the last goat path — using, as the Serbs would have been well aware, the very programmes used by the Nato planes that had bombed them back to the conference table. So, as one reporter put it, while the warlords were returning to their strongholds, the resort to virtual reality produced a virtual peace.

Actual reality produces its own gloss on all that. Travellers in the convoy carrying the Helsinki Citizens' Assembly's assorted idealists and observers to its annual conference in Tuzla got an insight into the nature of this Balkan war as soon as they reached Pocitelj, the first strategically placed small town on the Neretva river which was fiercely fought over by all three sides. All its bridges were blown for miles along, including the lovely, springing arch of the sixteenth century bridge of Mostar, a treasure lost for all Europe.

The pattern Pocitelj showed was to be repeated endlessly in towns and villages between Mostar, Tuzla and Sarajevo in the next fortnight. Yes, look, they *had* once all lived together side by side: here, three intact, ordinary village houses, with chickens and children playing in the yard, were followed immediately by two blackened ruins; another patch of normality — an old man snoozing by a doorstep — then more darkness, half a row this time, and so on; after which village life resumed, men in allotments straightening up to stare at the line of buses and white UNPROFOR jeeps. Further on, the devastation was more comprehensive — whole hamlets wiped out — until, in the Muslim part of Mostar, the coach was silenced by the extent of the wreckage and the improvised cemeteries of lately-dug graves in parks and roadside verges. For the scale of damage, only Sarajevo, at the end of my journey, seemed worse, probably because there it was the *modern* buildings of its wide, central boulevard — the familiar shapes of tower blocks and shopping centres — that were so broken and ravaged.

The harder part of this history lesson was about the specific character of the destruction. The village houses, especially, were not just ordinary, shell-damaged casualties of war; no, they had been dynamited, bulldozed, systematically razed. It was as though giant goblins had been on the rampage, stamping them flat. What added to the sense of falling into some terrible Balkan *Märchen* was that among these places were alpine villages scarcely touched by the twentieth century, high above the deep wooded valleys of central Bosnia: these archaic pastoral idylls, too were blackened with the fierce scorch marks of history. (The nineteenth century dealt with them no less savagely, of course. In despatches for the *Manchester Guardian*, Arthur Evans described the fires of destroyed villages

BOSNIAN DIARY

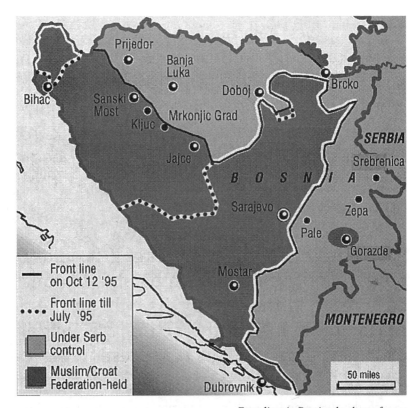

Front lines in Bosnia: the shape of peace

burning just as fiercely in the Balkan Wars of the 1870s; but then, of course, the valiant fighters for Free Bosnia were the Christian Serb peasants of the Krajina, struggling against the yoke of the Muslim landlord class, superior in strength and firepower...)

It started almost at the beginning of the war, and continued even beyond the end. A week after the ceasefire, Tim Garton Ash and Konstanty Gebert, the Polish journalist and aide to Tadeusz Mazowiecki (*Index* 5/1995), who had driven down from Zagreb through the Krajina, arrived in Tuzla half-dazed at the succession of empty, gutted Serb villages, still smouldering, which specially detailed Croatian loot-and-burn detachments had left in the wake of Operation Storm, whose name Franjo Tudjman so loves to pronounce. But weeks after the initialling of the agreement in Dayton, more than a month after the ceasefire, the fires were burning still near the Posavina corridor, where the Croats seem determined to hand back nothing but scorched earth.

These images compose eventually into a comprehensive diagram of the pathology of war, the fact that this is an unusual species of civil war being significant only because civil war is the extreme case, in which the Others who you must drive out, slaughter, eradicate are people known and seen in desperate close-up: your neighbours, even kin by marriage, made alien and fearful by the exciting madness you have been infected with. Not only must you get rid of them, you must make it impossible for them or theirs ever to return and resume what used to be normal life, for then your madness would be seen for what it is. So while it would have been easier if the goblins could have made the Muslims or the Serbs and their houses simply disappear, one does the best one can with dynamite and bulldozers. Obliterating the memory of a place — making lives and communities as though they had never been — must be one of the ultimate forms of censorship.

THERE IS an actual map which explains better than any narrative the complexity out of which all this came. This ethnographic map of Bosnia from the 1991 census must be one of the most extraordinary exhibits in the history of cartography, just as Bosnia's melting-pot was one of Europe's most curious pieces of nation-making. It shows at a glance the dense and complicated distribution of the Serb, Croat and Muslim mix across all the country's corrugation of mountains and valleys. Compare it with the maps of the ceasefire and what is known of the Dayton maps, and you see almost as quickly how few people are now at home in Bosnia. First, the Muslims of the eastern towns and villages above the Drina, whose centuries-old culture is recreated in the novels of Ivo Andric, were driven west. Then, late in the war, the Krajina (border) Serbs, descendants of those planted by the Habsburgs three centuries ago to stiffen the Croatian frontier against the Turk, were herded east by the US-backed Croat and Bosnian offensives of last May and August in the war's biggest ethnic cleansing — 200,000 or more driven into bulging Banja Luka and its hinterland — or replanted in unwelcoming Kosovo. And everywhere the skilled and educated from the towns had fled abroad, these economically viable ones much less likely to come back, of course, than the several hundred thousand whose permission to stay in makeshift asylum in Germany runs out, with Chancellor Kohl's patience, in March.

It will be some time before it becomes clear whether many displaced Bosnians will be able to go home, or whether the whole notion of restoring Bosnia's communities to their multicultural status quo ante, in which so much liberal hope has been invested, really is the naive illusion some US power brokers were always sure it was. In Sarajevo someone described Richard Holbrooke's explosive reaction to talk about the

 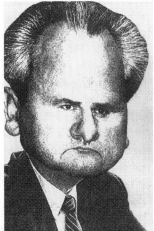 

Messrs Izetbegovic, Milosevic and Tudjman

importance of refugees returning to restore the uniquely mixed character of Bosnian communities: For Chrissake! They've got the whole map just about 90 per cent ethnically cleared. Don't start moving people around now and spoil it all: ie, don't upset this tidy new territorial balance which might get us some sort of settlement of these bloody Balkan quarrels.

Though no-one expects Muslims to return to Serb-held areas for years to come, the Bosniaks are otherwise committed to the return of refugees to their homes, and President Izetbegovic proposes his old city's tradition of tolerance and openness as the very pattern for democracy in Bosnia-Hercegovina: 'We do not see this as difficult. We are used to living like this... We were attending classes together with people who had different names, a different religion or nationality. None of this is strange to us.'

But other voices will tell you that that old Sarajevo is gone. Much of the old middle class, I was told, had evacuated itself smartly at the beginning of the war, leaving behind 'a militant mediocrity'. And Kris Janowski, the UNHCR's thoughtful spokesman, calculates that while it was true that 27 per cent of marriages in Bosnia before the war were ethnically mixed, there now remained only about 30 per cent of this original population, the rest being mostly refugees from the villages.

In Tuzla, that other celebrated example of the tradition Izetbegovic invoked, the ethnographic balance has been even more radically disturbed. Before the war, the 1991 census gave the proportion of Muslims to Serbs and Croats as roughly three to one; by 1995, UNHCR estimates showed, the tides of war had altered that proportion to 20 to one. The departure of many Serbs and Croats (though we saw some still in the surrounding villages, their churches undamaged), and the immense influx of Muslim refugees, leaving no billets for the hapless US troops whose headquarters the town has become, had made Tuzla statistically into a Muslim stronghold. In northern and eastern Bosnia-Hercegovina, the position is brutally reversed, the Muslim populations being reduced from 355,956 to 30,000 and 261,000 to 4,000 respectively.

The village houses, especially, were not just ordinary, shell-damaged casualties of war; they had been dynamited, bulldozed, systematically razed

Certainly the city's social democrat-dominated administration remains committed to Tuzla's tradition of tolerant multi-ethnicity. During Hitler's war, this miners' town's militant solidarity prevented the Nazis and their Croat *Ustashe* allies from taking its Jews to the death camps; and in October it played host for the second time to the hCa, drawing groups from all over Europe, not least a large contingent of opposition liberals from Serbia, who had taken three and a half days, via Hungary and Croatia, to complete a journey that used to take three hours from Belgrade. But survivors of the old Tuzla are well aware of what the changes may bring. Multi-ethnicity was organic to this place, says Sinan Alic, editor of one of the town's two independent newspapers, not some ideology imposed by the authorities; but what the new arrivals from the Muslim villages say to such old Tuzlaners when they question a narrowing of attitudes is: if you don't like the new situation, you can leave.

It's important to remember how much Bosnia was and is a peasant country; only a third of its population lived in the few towns of any size before the war. Revealingly, Karadzic, himself a village boy made good, if that's the word, blames the desire for a strong, multi-ethnic Bosnia he so deplores on 'certain pro-civic circles'. Displaced townsfolk, says a refugee administrator in one of the hCa's workshops, are generally much more confident about trying to return to their former homes, villagers far more scared.

UNSURPRISINGLY the census map was much cited as everyone counted their gains and their losses after Dayton. The Croats used it in arguments especially painful to them over the division (to give the Serbs a viable corridor) of Bosnian Posavina, along the valley of the Sava, and the future allocation of Brcko, postponed for a year when the agreement nearly came unstuck over it. This is one of the most dangerous minefields. Another, of course, is the reintegration of Sarajevo into Bosnia-Hercegovina, a question so vexed that President Chirac (Sarajevo is a French area of responsibility under the UN) has written to President Clinton saying that additional guarantees were needed for the 120,000 Serbs involved if the agreement was to be workable. 'The international community has become excessively involved in Sarajevo,' says Karadzic. 'It will have to pay the price, it will have to protect every single Serb house...for at least five years.'

Hardly less dangerous is the question of whether President Tudjman can — or cares to — make his Hercegovinian Croats work constructively with the Bosniaks in the Federation, and give up the brutal apartheid they enforced in places like Mostar. There the Muslims who live on the east bank, where 60 per cent of the buildings are damaged, were simply stopped by the Croat police from crossing to the city's administrative offices, installed in a hotel on the western, Croatian bank. In an attempt to make this bitterly divided place into a functioning city again, the European Union installed Hans Koschnick, the tough old ex-Oberbürgermeister of Bremen, as administrator, with a posse of policemen collected from several European countries. But the series of agreements he arduously negotiated, aimed at getting the Croats 'to stop stopping people', were repudiated almost as soon as they were initialled. If the new post-Dayton agreement is still working this month that will be a small miracle, and as good an augury for the Federation's future as anyone can hope for at present.

AT LEAST now there are no bodies in the rivers, no more mass graves being dug. For the long uneasy moment, there is peace. 'Cruel but realistic', as Muhamed Sacirbey described it; 'the only possible peace', in the words of his rather more satisfied Croatian opposite number, Mate Granic. Is what has been done a betrayal of Bosnia, as Denis Healey among others has suggested? At Tuzla, Timothy Garton Ash put it to Peter Galbraith, the US ambassador to Zagreb whom unkind persons call 'the tenth member of the Croatian cabinet', that what was in the making in the impending dealings was 'a Yalta for Bosnia'. But it will only be that if 'the international community' — which is to say Nato, which is to say the USA — lacks the will to do what the actual international community is to do through its common international institution, the United Nations: that is, whatever may be necessary to stop Croatia and Serbia taking Bosnia apart. (All the Balkan leaders, incidentally, and particularly Radovan Karadzic, address 'the international community' ad hominem, as if it were a surly bank manager, or a dodgy landlord who can't be trusted to keep the property in repair; or in the case of Franjo Tudjman, an indulgent rich uncle of whom he still has great expectations.)

If... The road ahead, as far as one can see through the snows of a Bosnian winter, is littered with 'ifs'. If implementation proceeds as quickly and firmly as President Izetbegovic insists it must, if it includes enough aid for reconstruction to make people think about the future rather than the past, if the US stays the course, and if no actual or metaphorical landmines blow the peace sky-high, then Bosnia may even manage something a little better than replacing unbearable misery with ordinary, everyday unhappiness, the therapist's traditional goal. But it will be some time yet before we can believe we have really got much further than the moment anticipated in the poem Laura Silber and Allan Little used as envoi to *The Death of Yugoslavia*:

> On that day we'll say to Hell: 'Have you had enough?'
> And Hell will answer: 'Is there more?' ❏

WL Webb *was formerly literary editor and assistant editor at the* Guardian, *and Guardian Research Fellow at Nuffield College, Oxford*

Unwanted occupation...

ON 28 JUNE 1989, at the site of the Field of Kosovo seen by Serbs as the cradle of their history, on its six-hundredth anniversary, Serbian President Slobodan Milosevic launched his nationalist *jihad* which led to the death of Yugoslavia and of so many of its citizens. He told a million Serbian pilgrims, inflamed with *slivovice* and corked-up patriotic ardour, to be prepared for battle. Since then, Kosovo has been the last old-style Eastern European police statelet, bristling with police and informers, its Albanian schools closed and political leaders imprisoned. The Albanians see it as an occupation. More than 300,000 of them have emigrated (*Index* 9/1992).

Apart from having its supporters demonstrate loudly outside the Dayton talks, the ethnic majority's response to the autumn planting in Kosovo of 9,000 unenthusiastic Krajina Serb refugees was to leak in their press the basic principles of the European Contact Group's 'Plan K' for Kosovo, said to include the restoration of the province's autonomy and an internationally supervised one-year transitional period during which discriminatory laws would be annulled, and democratic local elections prepared. The reports said that the plan would be discussed at the first conference on former Yugoslavia to be held after Dayton, but the stony US resistance to linkage between Kosovo's situation and the negotiation of Bosnia's ethnic dilemmas doesn't augur well for the plan's prospects. (Peter Galbraith, the US ambassador to Zagreb, known unkindly in some UN circles as 'the tenth member of the Croatian cabinet', bluntly told the Helsinki Citizens' Association's conference in Tuzla 10 days before the Dayton talks began that the question of civil rights in Kosovo and Sanjak were 'internal matters of Serbia and Montenegro'.) The day before the leak appeared, Astrit Saliu, a leading Kosovo Albanian journalist, was reported to have been arrested, beaten, and interrogated by the Serbian state security police. *WLW*

MELANIE FRIEND

...unwilling refugees

Photographs by Melanie Friend

PHOTO-FEATURE: KRAJINA REFUGEES

(Left, above) October 1995: Serbian refugees from Krajina, temporarily housed in the Boro-Ramiz Sports Centre, Pristina, Kosovo
(Left, below) October 1995: Serbian soldier from Krajina, Boro-Ramiz Sports Centre
(Above) October 1995: Visnja Samardzija from Krajina, Boro-Ramiz Sports Centre

(Above) Centre for Anti-War Action, Belgrade, September 1995: Dragana Roknic from Krajina tells her story to a counsellor while registering as a refugee
(Below) Boro-Ramis Sports Centre, Pristina, Kosovo, October 1995: Jelka Bekic (63) comforts Anka Skrgic (85), both refugees from Krajina

hCa's campaign to support independent media in the territories of the former Yugoslavia

THE DAYTON accords suggest a fragile 'peace' may now be implemented in Bosnia-Hercegovina. If this fragile peace is to have any hope of bringing long-term stability to Bosnia-Hercegovina and the territories of the former Yugoslavia (TFY) in general civic groups, independent media and all others working for reconciliation and dialogue need our support now.

Independent media will continue to play a key role in overcoming the divisions that are the result of four years of war and suffering. At the moment independent media are under-resourced and overstretched in every sense, journalists continue to suffer from political and other pressure, and many newspapers, radio and TV stations continue to operate in an isolated environment. Their situation will not improve as a result of the Dayton settlement, if anything it will deteriorate.

As a part of its work to support the civic and economic regeneration of Bosnia-Hercegovina and the TFY in general through partnerships and twinnings, the Helsinki Citizens Assembly is launching an international campaign in support of independent media. The goal is simple: to generate long-term support for the independent media through direct partnerships or twinnings with newspapers and radio/TV stations from outside the region. Independent media from all over the territories of the former Yugoslavia desperately need to widen their contacts with independent media from elsewhere.

The rewards of this campaign will be
- to widen the support base of independent media;
- to generate direct links with newspapers/radio/TV stations from outside the region, encouraging professional and expert assistance to overcome their isolation;
- to establish a new support network within the hCa which may be used to channel new funds into independent media.

If you think you can help in any way, contact Julia Glyn-Pickett, hCa, c/o **Index on Censorship**, *33 Islington High Street, London N1 9LH*

LETTER FROM JERUSALEM

STANLEY COHEN

The kitsch of death and peace

A MONTH AFTER Yitzhak Rabin's assassination on 4 November — and daily life in Israel is only slowly emerging from living in a CNN news item. For a long time, every hysterical cliché seemed to come closer to this simulation; now the older realities are asserting themselves.

Every day still brings new dramas. More revelations about the tangled web of right-wing religious organisations in which the assassin was immersed. More details about the bunglings of the security services. More self-flagellation — some genuine, more sickeningly bogus — from 'moderate' religious and right-wing figures about where they had gone wrong. And now a meta-debate about there being enough, or else too much, self-flagellation.

Images of the initial death kitsch still remain: the memorial candles, wreaths, songs and poems; the hundreds of thousands of pilgrims flowing to the three shrines of the grave, Tel Aviv square where the shooting took place and the Rabin family's Tel Aviv apartment; delegations of schoolchildren reciting their poems and showing their paintings of crying doves; teenagers in army uniform huddled sobbing in the streets; the endless sanctifying tributes in the media; the iconography of Leah Rabin as Jackie Kennedy-Evita Peron.

With tears in their eyes, people appeared on the TV screen to ask: 'What's become of us?' They said that they were now living in a different country; nothing would ever be the same. They claimed to feel that they had lost their father (psychoanalytic allusions to the primal horde became banal). They expressed guilt: 'We should have supported him when he was alive.' They anguished about 'Jew killing Jew', brother being set against brother (there are no sisters in the public discourse) and the need for national reunification, cleansing and soul-searching.

'Spontaneous outpourings of grief.' If ever the journalistic cliché applies, then this was it. Not even the most cynical outside observer could compare these scenes to the mourning for Stalin, Mao or Kim Il Sung. Nothing was orchestrated or enforced. Indeed, if one meaning of the term 'kitsch' is that fake sentiments are packaged and demonstrated as if they were real, then there was nothing kitsch about any of this. There was genuine grief; people seemed wholly identified with the public spectacle. On Clinton's lips, the slogan '*Shalom, chaver*' (goodbye/peace, friend) — scripted for his initial press conference and repeated at the graveside — was pure Disneyland. Its instant adoption in Israel however — first in improvised scribbles, now as mass-produced bumper stickers and T-shirts — expressed a real sense of loss.

Unlike the Kennedy assassination, there was no obscurity about the assassin's aims. Nor was American society in 1963 so obviously divided as Israel today. Everybody understands the meaning of Rabin's killing — and still does, despite the current murky theories about moles, agents provocateurs and fake organisations. This must be one of the purest cases of political assassination in history. Yigal Amir could not be dismissed as a lone autistic who wanted a moment of fame. He acted rationally and with organisational subcultural support. He served in an elite combat unit and was a good final-year law student (albeit at Bar-Ilan, the oxymoronic 'religious university'). Far from being on the margins, Amir came from the epicentre of that large segment of a generation of Israeli youth educated within the confines of fundamentalist Jewish nationalism.

Amir and his associates (six more, including his brother have now been charged with various degrees of collusion) needed no occult indoctrination in the closed worlds of army *yeshivas*, or terrorist cells to pick up their justifications. Every strand in the belief system — deification of the land, racism, contempt for democracy, the will of God as transcending secular law — was quite open in the wider political culture. The graffiti ('Rabin the Traitor') were on the walls; car stickers read 'Rabin and Peres First' (alluding to 'Gaza and Jericho First'); placards at demonstrations showed Rabin in SS uniform; the rabbis openly preached and issued their public *fatwas*; the ideology was casually recycled on TV. (Instead of Oprah Winfrey confessions about incest and sexual abuse, popular Israeli TV talk-shows will host a settler who nonchalantly says that if his own son were among soldiers sent to evacuate a settlement, he would open fire at him as well.)

There was, however, one crucial way in which Amir and his friends appeared to mis-cue their reading of Israeli society. His most chilling reported words (so far, at least) were: 'Before this, I thought everyone was on my side.' Living within their self-referential, ideological universe, they could not have anticipated that the assassination would open up such a sentimental wave for 'peace' or would allow Rabin — an unlikely such figure in his lifetime — to be reincarnated as a populist martyr for peace.

However rationally the event's origins can be explained, it took on mythic proportions beyond the rational. Every textual allusion to the Bible, Greek tragedy or Shakespeare was made. The man of war receives the Nobel peace prize; swords turn into ploughshares. The icons of John Kennedy, Abraham Lincoln and even Martin Luther King were evoked. Not to mention Moses who lead his people to the edge of the promised land but did not live long enough to see the valley beyond. And Isaac (Yitzhak) who is sacrificed on the altar of peace. Eitan Haber, Rabin's political aide, ended his eulogy at the grave by reading from the bloodstained paper on which the words of the 'Song of Peace' were written. As he left the mass peace rally, Rabin had folded this page and placed it in his pocket, just near the bullet wounds. Not even Hollywood would have dared to script this.

But it happened and no amount of ironical textual deconstruction will remove these events nor erase the sentiments they evoked. Still, we have to take a deep breath and remind ourselves of three other realities that have not appeared on CNN.

The first is that this frenzied reinforcement of primal tribal unity is a little suspect. The profound divisions within the historical consensus remain. This is not just a matter of being for or against the Oslo agreements, nor even religious versus secular. At stake is the contradiction between ethnic nationalism and democracy. The right has now recovered its poise. The new-found zeal of the police and attorney-general's office to prosecute the slightest hints of 'incitement' (a silly over-reaction to their blatant tolerance of right-wing and settler violence for years) has given the right the rallying call of 'witch-hunt'. They are now projecting themselves as the victims of secular liberal intolerance.

LETTER FROM JERUSALEM

Long ago in Oslo: Rabin and Arafat receive 1994 Nobel Peace Prize

And even the defenders of light against the forces of darkness have shown their unacknowledged ethnic exclusivity. For them too, 'Jew killing Jew' was the ultimate taboo. But suddenly, the values of tolerance, the rule of law and democracy became proclaimed as if they always had been the core identifying features of Israeli society, the moral centre from which this act was a horrible aberration. This is a remarkable epic of instant historical revisionism which allows the Labour Party to disavow all responsibility for its entire past record of violence and injustice. Life before Oslo goes down the black hole of memory. So too does the capitulation on human rights issues by *Meretz* (the liberal doves) during its three years in the coalition.

The second missing reality is another piece of historical revisionism. The dominant Clinton-rhetoric avoids the consistencies in Rabin's biography that are as impressive as any putative ideological conversion he is supposed to have undergone in the last three years. Far from being the 'traitor' of the right's denunciation, he remained faithful to Labour Party Zionism. In 1948, he carried out Ben-Gurion's tacit orders to forcibly expel the Arab population of Lod and Ramle. Forty-five years later, during the 1992 election campaign, the same Rabin would respond to right-wing hecklers demanding 'transfer' (the Israeli version of ethnic-cleansing): 'Don't talk to me about transfer, no-one in this hall has transferred more Arabs than I have.' This was the same Rabin who 'advised' Sharon during the siege of Beirut; who gave orders during the *intifada* to break Palestinians' legs and arms; who sanctioned the army's undercover death squads; who up to the end, showed utter contempt for the language of human rights.

The Oslo agreement is manifestly unjust to the Palestinians. They had to accept a very poor deal... without territorial integrity or national sovereignty

We all understand that in terms of pure political power — external consequences rather than inner conversions — none of this might matter at all. The conventional wisdom is correct: precisely because of who he was and what he stood for, only Rabin had the authority to break so irrevocably from the deadly, wasted decade of Likud governments. He might not exactly have been a de Klerk or de Gaulle, but looking back — rather than how he looked at the time — these historical comparisons will be earned. Whether Shimon Peres can capitalise on the 'give peace a chance' sentiment is unclear. His task, though, would be easier if Rabin had only shown some of the spirit of reconciliation that Clinton-kitsch attributes to him.

This leads to the third and far most important political reality absent from the month of mourning. The simple fact is that the Oslo agreement — and even more, its grudging implementation in Stage One and Two — is manifestly unjust to the Palestinians. They had to accept a very poor deal, an arrangement which redeploys the Occupation by placing them in administrative control over only 30 per cent of the West Bank, without territorial integrity or national sovereignty. On every major issue — land, water, settlements, Jerusalem, refugee rights — Palestinian aspirations have either been blocked or put into limbo. Above all, there is the shameless continuation of land expropriation and settlement construction. Meanwhile Arafat's authoritarian regime tightens its grip — intolerance of criticism, torture, censorship, a network of security forces and informers, corruption, a special court system more unjust than Israeli military courts.

The Israeli sense that nothing will be the same again is true. Yet it is also an expression of the profound narcissism of Israeli political culture. Outside, everything is the same as it was before the assassination: the road-blocks; the crazy map of Areas A, B and C (designating degrees of Palestinian autonomy); the unrepentant hard core of the settlers; Hamas and Islamic Jihad terrorism; and the democratic Palestinian opposition.

The Israeli political leadership faces the same decisions as before the assassination. The choice is not between reversing or continuing Rabin's direction; his achievement was that it is too late to go back. The choice is whether to convert the easy kitsch of peace into the hard concessions of justice. Anything less would be a cynical manipulation of the genuine and moving populist sentiments which the first wave of mourning revealed. These emotions remain: polls now show that nearly 75 per cent of the population favour continuing negotiations with the Palestinians.

The willingness of the majority of the public to make some compromises is real, but volatile. As the Israeli leftist journal, *The Other Front*, rightly argued: 'The rise and fall of professed support for compromise is largely a function of the government determination, or lack of it, to press forward.' The zeal in creating fake internal reconciliation would be better spent in confronting the radical — and therefore divisive — concessions needed for reconciliation with the Palestinians. This, after all, is what the whole conflict is about. ❑

Stanley Cohen *is professor of criminology at the Hebrew University, Jerusalem. He has just been appointed Martin White professor of sociology at the London School of Economics*

IN THE NEWS

IRENA MARYNIAK

Notes from prison

When they visited me in prison they would always emphasise that the case was in their hands, that the judges, the prosecution, the whole system was subordinate to them. Nobody can live on a salary, they said. This is the free market. Everyone has to find other means to survive: bribes, business interests, protection rackets... They were pretty convincing about their unlimited power. When I asked them about the practicalities of controlling the judiciary, the two FSS men rocked with laughter. First you have to 'work' on them, they volunteered cheerfully, catch them out when they're not on the level. Then you take them aside and paint their future in the darkest possible terms — if they're exposed. By then they'll be sweating and ready to do anything to ensure that the FSS preserve their reputation. Then you sweeten them up. After that they'll be only too keen to show their gratitude. They'll be under threat of compromise for life.

It had taken me a long time to understand why the two FSS men had kept so 'sweetly' persuading me to give them information about drug-taking among children of politicians, financiers, and the artistic elite. They laughed, as ever. 'It's dynamite,' they said. 'You can get anyone by the throat if you threaten to compromise their children. After all,' they said, amazed at my inability to understand these elementary things, 'the elections are coming.'

Alina Vitukhnovskaya, *Notes from Butyrsk Prison*, 1995

THE RUSSIAN parliamentary elections are upon us as *Index* goes to press, and 22-year-old poet Alina Vitukhnovskaya is out on bail. It is now more than 15 months since she was imprisoned and charged with involvement in a drugs racket which, it has been alleged, threatens to undermine the very foundations of Russian society.

Evidence to support the charges is flimsy, to say the least. And it comes direct from the Federal Security Service (FSS), the latest incarnation of the KGB which, a few years ago, engineered the imprisonment of political dissidents Konstantin Azadovsky and Yuri Edelshtein on similar charges. If convicted, Vitukhnovskaya could spend up to 15 years in prison. For the time being she is back home following a hearing on 28 November at which no barrister appeared, and which came to an abrupt end when someone announced that a bomb had been planted in the courtroom.

Vitukhnovskaya is reputed to be one of the most talented and distinguished young poets of her generation. On the night of 16 October 1994, she was arrested by a dozen or so FSS men who were waiting for her outside her block of flats as she came back from a rock concert. They charged her with pushing two grams of LSD to two youths at a Moscow metro station. The evidence against her lies in the money — equivalent to US$20 — which she was apparently carrying when she was arrested, and two vials of drugs said to have been found in a speedily conducted search of her room. The main prosecution witnesses are two young men who picked Alina out of a solo identity parade conducted by the police. They are reported to have been high at the time.

She was held in Butyrsk prison for a year without trial. Her sight and hearing have deteriorated; she has frequent fainting fits; she is said to have been close to suicide. Prison conditions in Russia are abject. 'Rats' tails in the soup,' she says impassively. 'Sometimes I wouldn't eat for days.' In June, Russian PEN took up her case and appointed four prominent writers to follow the proceedings and support her defence. An extraordinary picture of the Russian judicial and penal system has emerged.

Court hearings have been erratic and inconsistent, witnesses bewildered and uncertain of their evidence. Signs of police coaching have been blatant; stories have failed to match. Signatures were forged, documents back-dated and journalists sent home on the grounds that hearings had been postponed, minutes before they actually began. The case for the defence has been ignored throughout.

The FSS began to take an interest in Vitukhnovskaya following publication of an article in the journal *Novoye Vremya*, in which she gave an original, personal account of her encounters with Moscow's alternative youth culture and its perception of drug-taking. Her telephone was tapped, it seems, and later she was invited — unsuccessfully — to inform on children of prominent political and business figures.

Vitukhnovskaya's lawyer, Karen Nersisyan, has called her case a gauge of public mood before the elections. It is also a test of the power wielded by the FSS and the police. 'There are tens of thousands of cases like this in Russian courts at this moment,' Nersisyan says. 'People are being charged on the basis of suborned witnesses and a couple of bits of paper.' In police cells, intimidation and torture are rife. In 1995, 900 police officers were dismissed for abuse of power or corruption in Moscow alone; a further 600 face disciplinary proceedings. Victims are generally unidentified.

'I suppose I can gain some aesthetic satisfaction from the way things have turned out,' Vitukhnovskaya says, her voice strangely flat and serene and close down the phone line. 'But it may be more dangerous now. This isn't freedom. The leadership we have here is bound not by law, duty or responsibility, but by an exclusive kind of insanity. I feel I'm being dragged down by an invisible slipstream, along with everyone else who happens to have been caught up in it with me.' ❑

ISABEL HILTON

Tibetan succession

China puts up its rural candidate for Panchen Lama and faces Tibet's religious leaders — with a hard choice

A PURGE of the Tibetan religious establishment appears to be in preparation following the intervention of the Chinese government in one of Tibet's most sacred religious procedures: the selection of the eleventh Panchen Lama.

Amid a dramatic tightening of security in Tibet, the Chinese government forced Tibetan religious leaders to reject the choice of the exiled Dalai Lama, the six-year-old Gedhun Choekyi Nyima, as the reincarnation of the tenth Panchen Lama and to nominate another child through a lottery.

The Chinese crackdown on Tibetan religious freedom has been accompanied by an intense propaganda campaign against the Dalai Lama, who is still venerated by most Tibetans as their supreme religious leader. He has been accused by the Chinese government of violating religious principles and seeking 'to split the motherland'. Tibetans have been warned that in the practice of their religion, 'patriotism', by which Beijing means loyalty to the Chinese government, must come first. The crisis marks an end to a period of relative religious freedom in Tibet.

The controversy over the selection of the Panchen Lama's reincarnation has been implicit since the death of the tenth incarnation in Shigatse in 1989. The late Panchen Lama had been a controversial figure for both sides in this dispute, but on his death was hailed by both sides as an important leader. For the Tibetans, he was recognised in his later years as a man who struggled to preserve and restore Tibetan cultural and religious traditions after the devastation wrought on the country since the late '50s. Beijing, on the other hand, paid tribute to his 'patriotism' and called him a great leader of the Tibetan people. The task of locating his reincarnation was given to a committee headed by the abbot of the Panchen Lama's historic seat, Tashilhunpo monastery, Chadrel Rinpoche. But for Chadrel, as a devoted member of the Gelupga sect, the only person with the right finally to choose was the Dalai Lama. Despite repeated pleas from both Chadrel and the Dalai Lama, the Chinese government has refused to admit the Dalai Lama to the process.

In May of this year, the Dalai Lama proclaimed his choice, after information on a number of candidates was smuggled to India. His proclamation provoked a furious reaction in Beijing and Chadrel Rinpoche was detained. He has not been seen since. The child, Gedhun Choekyi Nyima and his family also vanished and have been reported to be in Chinese custody in Beijing. Despite what appears to have been the strenuous resistance of the most respected figures of Tibetan Buddhism, the government last month forced the rejection of the Dalai Lama's candidate. It was, they said, a matter of sovereignty. On 29 November, a ceremony was held in the Jokhang Temple in Lhasa in which the name of another six-year-old boy, Gyaltsen Norbu, was selected. His name was one of three that had been placed in the Golden Urn, a vessel sent to Tibet in the Qing dynasty to serve as the vessel for a lottery. It was intended to symbolise the Qing emperors' overlordship of Tibet, symbolism that has been co-opted by the present government. Gyaltsen Norbu was formally installed in Tashilhunpo on 8 December 1995.

For Tibetans, this is a painful setback to what had been a gradual, if limited, restoration of certain religious freedoms. The Panchen Lama is, after the Dalai Lama, the second most revered religious authority in Tibetan Buddhism and the only one, apart from the Dalai Lama, who commands national allegiance. Traditionally the Dalai Lama and the Panchen Lama have exchanged the roles of master and pupil and, on the decease of one, the other has been acknowledged as the most important religious authority. Because the two great lamas were traditionally so close, each has felt a special responsibility in the critical search for the reincarnation of the other.

For practising Buddhists to be forced to reject the Dalai Lama's candidate is more than painful. To disobey the highest religious authority is also a transgression against faith. Many clearly resisted: more than 48 people have been arrested since May 1995 for their loyalty to the Dalai Lama's choice, including more than 30 monks from Tashilhunpo monastery. The principal detainee is, of course, the abbot himself, Chadrel Rinpoche, against whom a campaign has now been launched that is couched in language not seen since the Cultural Revolution. Tibetan religious leaders have been told to prepare criticisms of the Dalai Lama and Chadrel Rinpoche. Their loyalty and patriotism is to be judged by these statements. There have been signs for several months of a tightening of security in Tibet amid admissions that Tibetan sympathy for 'splittists' — by which the Chinese mean the Dalai Lama — remains prevalent. The Chinese leadership has intervened at the highest level in this religious procedure: Jiang Zemin and other Party leaders addressed a meeting of Tibetan religious leaders in mid-November 1995, at which the rejection of the Dalai Lama's candidate was finally announced. Since the Tibetan people are unlikely to accept a Panchen Lama who has been imposed by the Chinese government, there are fears for the safety and future of Gedhun Choekyi Nyima. The Dalai Lama has appealed to the Chinese leadership to permit the boy to have a religious education, but the government is unlikely to allow him to reappear for fear of the devotion that he would inspire. ❑

Isabel Hilton is a writer and broadcaster. She is currently writing a book on the Panchen Lama

KARIM ALRAWI

Military sidestep

Egypt's elections were a farce, with voters squeezed between government, fundamentalists and the army

IN SEPTEMBER 1995 the Egyptian Organization for Human Rights (EOHR) published its second full report on freedom of opinion and expression in Egypt. An interim report had been published in 1993 which expressed serious concerns about the deteriorating situation in the country. It noted instances of the recruitment of academics and journalists by both the government and the religious fundamentalists to exert pressure on writers to take sides in the slow, simmering civil war that is affecting all sectors of society. The new report, entitled Silenced Voices, is full of instances of the use of the legal system to subvert the freedom of writers and academics to express their views and disseminate opinions that are contrary to those of either the

---

Do you want to disappear?

Wipe the slate clean? Cover your tracks? TV company researching for a Channel 4 documentary needs engaging individuals to go through the process of pursuing a new life and starting over again, thus exposing how much information is kept on us.

Contact: 0171 354 0052

government or the Islamic opposition.

The EOHR remains one of the few genuinely independent organisations of high integrity reporting on events in the country. As would be expected, there is no discernible bias in its report: the EOHR takes both sides to task equally for the deteriorating and depressing situation. On the one hand are the government's anti-democratic workings, witnessed once again by the coercion of opposition parties during the latest parliamentary elections on 29 November; on the other, the intolerance of the religious groups to any point of view other than their own. Despite their bitter dislike for each other, both parties seem to reserve their greatest hostility for the independent, or free-thinking, writer. The financially corrupt government of President Hosni Mubarak fears the call to accountability that is implicit in the demand for democratic reforms more than it fears the death threats of the Islamic groups, who, in their turn, view democratic and secular ideas as a threat to their own authoritarian, God-given right to govern Egypt in a manner most of its population abhor.

The report describes how Islamic lawyers are forcing the law courts to adopt medieval concepts. One such case is that of Dr Nasre Abu Zaid, whose separation from his wife Dr Ebtihal Younis it ordered, on the dubious charge of his unproven apostasy. This has also resulted in Dr Abu Zaid's position at the university being frozen until an appeal has been completed (*Index* 1&2/1994).

The government in its turn has chosen to side-step the law courts completely, bringing journalists and writers, as well as political opponents, before military tribunals. There is no appeal except to the president himself; he has shown himself to be singularly unmoved by the brutal and summary nature of military justice. As a result of the undermining of the legal system by the further encroachment of the military on civilian life, Egypt has experienced a wave of executions unprecedented in its modern history. In the first nine months of 1994 at least 22 civilians were sentenced to death, according to the US State Department. Such power over the legal system has also led the security forces to other systematic abuses. Extrajudicial killings are now a regular feature of life: state security men gun down suspects in city streets or in the countryside. The EOHR has a list of such incidents but admits that the scale of the killings may be much greater than they have so far recorded. The Islamists respond in similar fashion, killing policemen and conscripts as well as officers.

Torture is routine. The victims include women and children as young as 12 and 13. Such reports of brutality are corroborated in the US State Department's own *Human Rights Report on Egypt* (1995). The irony of such a report by the US government should not go unnoted, as several of the officers supervising torture in a systematic way were themselves trained in these techniques at MacArthur Airforce Base in Florida.

Given a situation where the freedom to live any kind of a life is severely curtailed, it is difficult to isolate freedom of expression. Nevertheless, the EOHR puts on a brave face and systematically points out in its report the laws contravened and the clauses in the Constitution that are breached by government abuse. They quote chapter and verse in defence of the rule of law in a country where neither government nor opposition respect any law. President Mubarak suspended the Constitution when he first took office in 1981. The protection of the legislated laws of the land has been overruled by the anti-terrorist Emergency Law which is in effect until 31 May 1997, and can be renewed indefinitely.

The inability of the law to protect its own was nowhere more evident than in the security forces' arrest of 41 lawyers protesting the death in detention of their colleague Abdelarith Madani, in April 1994. Most were held without charge or trial for three to four weeks. This is in addition to about 150 former defendants who have received release orders from the courts and are still being detained illegally, some for almost four years without further charge or trial. In some cases their lawyers are also being detained.

The government's general lack of concern for the welfare of its citizens was most recently demonstrated by its willingness to condone the savage and brutal practice of female genital mutilation (FGM). This un-Islamic and un-Christian custom is carried out in both communities. According to the government's own figures, between 70 and 80 per cent of Egyptian women undergo this ritual mutilation during childhood. Instead of trying to raise awareness to combat this cruel practice through educational and health programmes, the Ministry of Health decided in October 1994 to permit the use of government hospitals and allow medical staff to carry out this mutilation. To its credit, the EOHR has taken up the challenge of trying to shame the government into withdrawing its effective endorsement of FGM. They have also taken legal action against the senior cleric at Al-Azhar, Egypt's Islamic University, for his statements in support of the practice. Their action places them, literally, in the firing line; Sheikh Gad el-Haz is both a government official as well as a fundamentalist sympathiser. Legal action against him means that the EOHR is no longer just a witness to the criminality of both the government of President Hosni Mubarak and the Islamic gangs, but has entered the fray as a player, albeit on the side of sanity. It remains to be seen whether it can survive the encounter. ❑

Karim Alrawi *is an Egyptian writer and dramatist now living in Canada*

Communicating human rights

Benaco camp, Tanzania: home to 350,000 Rwandan refugees

Reports of human rights abuses proliferate, yet violations continue on an overwhelming scale. A meeting of human rights groups in Oxford asks the question: where are we going wrong?

STANLEY COHEN

Witnessing the truth

'A camera in the right hands at the right time in the right place can be more powerful than tanks and guns'
Peter Gabriel announcing the 'Witness Program'

IN 1992, supported by a grant from the Reebok Foundation, the New York-based international human rights organisation, the Committee for Human Rights, launched its 'Witness Program'. The programme distributes video cameras to hundreds of human rights observers throughout the world and trains them to record violations as they occur. Governments should no longer be able to deny the facts. International audiences can be shown not just words, claims and allegations but vivid pictures: a torture victim's scars, soldiers shooting a peaceful demonstrator.

The international media regime demands visual imagery and human rights organisations should try any ethical methods to supplement their traditional written information. But behind these technological innovations in recording information (and the use of electronic mail to increase its flow) there lies a very old-fashioned faith in the power of all this knowledge. The human rights movement is one of the few survivors of the Enlightenment project. It not only upholds the ideal of universal values and standards, but also assumes that information about their violations will produce universal moral and emotional responses. The power of the camera just extends the power of the word.

Moral witness might be a value in itself; knowledge might indeed speak (electronically now) to power. But the human rights movement must also be sustained by the more utilitarian belief that communicating this information will actually have some effect on our audiences. If only people could know (or see) what is going on — then they would wake up, react, do something.

It is just this faith that is contradicted by the daily practice of human rights work. The massive amount of information we generate is not matched by public responses appropriate to the scale of the atrocities. Nor are the resources devoted to collecting this information matched by sustained thinking about how it should be disseminated or what effect it might have on its intended audiences. Reports are 'released' — without evaluating or even monitoring their impact. A sophisticated international system of moral bookkeeping carefully collects, checks, processes and circulates information about violations all over the world. But we have little idea what this information does to people. Sitting in your home in New York, London or Paris, what does it mean to know that another 5,000 people have been slaughtered in Rwanda, that 20 Kurdish detainees have been tortured in Turkey, that 100 more street children have been killed in Bogotá?

Most human rights organisations are not interested in such questions because they are not explicitly trying to evoke reactions from the general public. Their information circulates in a closed circuit made up of other like-minded organisations, governments, United Nations and other inter-governmental agencies, lawyers and academics. Within this circuit, the information is indeed self-evident, the facts do speak for themselves. For the more sympathetic parts of this audience, these are more details to be assimilated into a familiar cognitive map: further items of news about what's happening in Zaire, Syria or Peru. For target governments, their protectors or allies, this is just more information to be ignored, denied, explained away or justified.

Much human rights information remains in this loop, endlessly circulating in reports, self-referential official documents, fact-finding missions, commissions, enquiries, UN and legal texts. It simply does not reach a wider public. And when it does, it drops into some kind of cognitive black hole. The information itself is not denied any more than you can 'deny' that there are homeless people begging on the streets of London. But you accommodate to knowledge of atrocities ('that's the sort of thing that happens in places like that') or just feel helpless ('what can I do about things like that?').

This happens even to the dramatic stories that pass the media threshold — Bosnia, Somalia, Rwanda, Burundi, Haiti. The initial attention, compassion and outrage seem to burn themselves out. 'I can't get those horrible pictures of corpses out of my mind.' But most of us eventually do. We cannot easily or indefinitely live with this type of information. So it becomes lost — preserved and recycled in yet another Amnesty or Human Rights Watch 'Follow-up Report' which records that indeed the same old things are still happening in places like that. Other information seldom even leaves the internal loop: the 'Chad Rule' is that no-one wants to hear about places like Chad.

FOR THE LAST two years, I have been studying how major international human rights organisations disseminate their information to the wider public and how to evaluate the effect of this information. This project grew out of my experience in human rights work in Israel and the Occupied Territories.

This is a rather special case for human rights reporters. On the one hand, the violations against the Palestinians during Israel's 28-year military occupation are characteristic of any such coercive regime. Torture, administrative detention, extrajudicial execution, house demolition, restrictions on movement, deportations...this is all predictable. On the other hand, these abuses happened (and, in the shadow of the 'Peace Process', still happen on a more restricted scale) while Israeli society itself has retained all the formal contours of democracy. Freedom of expression and assembly, the rule of law and academic freedom have all remained intact.

Unlike most societies where gross human rights violations are routine, this relative openness means that knowledge of what is happening is available and accessible. Investigative journalism (which blossomed over the *intifada* years since 1988) regularly provides the Israeli public with atrocity news. With a few rare exceptions, organisations like *Index on Censorship* or Article 19 are not needed to protect Israeli or international journalists. And — as the Israeli government correctly replies to its critics — human rights organisations, whether Israeli, international and even Palestinian, benefit from an access and freedom to report quite unthinkable in repressive societies elsewhere in the Middle East. On one subject alone — allegations of torture and mistreatment of Palestinian detainees — five major reports were produced and widely circulated over four years. These were all composed of detailed, vivid and irrefutable evidence.

No Israelis, then, can say in good faith that 'we didn't know' nor claim that there were restrictions on their freedom and opportunity to protest. Yet the level of reaction — even within the dwindling liberal sectors of the population — has always seemed less than this cumulative information 'should' produce. There are good and bad reasons for this gap — rooted in the history of Zionism, the nature of the Israeli-Palestinian conflict, Jewish identity, the real threat of terrorism, the sense of permanent insecurity, patriotic kitsch and unacknowledged racism. This is the unique political culture in which an Israeli, Palestinian or international human rights organisation disseminates its reports.

This gap between knowledge and action — what you know and what you do — is not, of course, unique to Israel. Human rights information is always selectively filtered — whether it is easily available or whether it has to penetrate the densest official censorship, intimidation, lies and propaganda. This filter is the obstacle to all forms of political mobilisation. Every social movement — feminism, environmentalism, animal rights — tries to use information (their own variants of atrocity stories) to arouse appropriate responses. In another sense, the problem is no different from any form of marketing, education or persuasion: why does the obvious information about the dangers of tobacco not lead everyone to give up smoking?

But human rights organisations seldom see things this way. Their discourse is dominated by lawyers, the monopoly of a specialised professional elite. They hardly regard themselves as being in the business of mass communication. They resist seeing their information as a product or a means of persuasion. The notable exception is Amnesty International, where the monitoring and interventionist work of the International Secretariat is used by national sections to build a participatory social movement.

Even the most elite legalistic organisations, though, are aware of the problem of getting their reports off the shelves. There are three different routes through which human rights information flows.

First, there are national organisations working primarily in their own domestic settings. This covers a wide range. On the one hand, there are countries (like my Israeli case) where the information is publicly accessible. On the other, there are the many variants of repressive states where information cannot be collected or internally disseminated. Censorship, political closure and the culture of fear do not allow public acknowledgement of what is privately known. In Havel's memorable phrase, people 'live in the lie'.

Second, there is the work of national organisations trying to send information into the international arena — whether to international NGOs, UN enforcement agencies or the global media. Here the problems obviously go beyond the technical solutions of e-mail or video footage. Communication depends on geopolitical interests and the agenda setting of the new world order. There is also the dependency which make a single story in the *New York Times* or an item on CNN more valuable than any amount of domestic attention.

Third, there is the network of international organisations dealing with violations that happen in distant places. The political problems here are familiar: how to enforce universal human rights standards, mobilising international pressure, humanitarian intervention. Far less attention has been paid to how the basic commodity of human rights work — the information that bad things are happening out there — is communicated. Who gets to know this information? What impact does it have?

Within the labyrinth of the international human rights enforcement apparatus, communication takes place through a formal discourse of covenants, treaties, conventions, declarations, reports, protocols and resolutions. Work is carried out by officials, experts, professionals and bureaucrats. Their particular style of communication is 'UN-speak': bland, technocratic, legalistic and designed not to offend. The language is impenetrable to those outside the loop — often so abstract, non-pictorial and non-specific that it is unclear what exactly is being talked about. Victims do not seem to appear on the agenda; names of perpetrator countries ('state parties') are sometimes not even mentioned; a country five years late in submitting a report to a committee will gently be reminded to be on time next year.

This information is self-referential in the sense that its impact is measured in terms of criteria generated within the loop itself. Has this report led to a special rapporteur being appointed? Has a working party been set up to 'consider the applicability of the Optional Protocol to the Covenant'? Has a right designated as 'civil' been moved to a list consisting of 'social and economic' rights? Of course such decisions may have important consequences. But it is unclear, even to the most informed readership, just what these paper manoeuvres mean.

All social problems have owners and managers — and the human rights problem has come to be officially owned and managed by a family

HUMAN RIGHTS: WITNESS

Bogota, Colombia: one of Latin America's vast army of street children

partnership of lawyers and international bureaucrats. The international community's success in responding to the horrors of our century by creating universal standards and agencies of prevention is undeniable. But it has created new barriers to public access. Both the pictorial language of atrocity, horror and suffering and the older theological language of sin and evil have given way to legalism and UN-speak. Not 'Thou shalt not kill,' but 'Thou shalt not violate the right to life protected by Article 6, Paragraph 1 of the Covenant.'

THE ACTIVIST and monitoring NGOs on the edges or right outside this official circuit are more interested in reaching a wider public. The reports by international organisations like Human Rights Watch, Lawyers' Committee for Human Rights and Amnesty International have generated an extraordinary amount of information over the last 30 years. It would be difficult to find any item in the catalogue of violations, any corner of the globe, that has not been exposed: religious repression in China, attacks on press freedom in Kenya, genital mutilation in the Sudan, police killing of street children in Brazil, child labour in Kashmir, land mines in Angola, torture in Turkey, political massacres in East Timor.

Unlike the proprietors of the official human rights business, the activists in these organisations are genuinely concerned about what happens to all this information. Their reports try to balance professional dialect (for example, references to legal standards, phrases such as 'We are deeply concerned about...') and vivid, accessible descriptions (for example, concrete personal cases of victims). But they are often difficult to read; even people within the community confess that they have never been able to get through a whole report.

The primary way to reach the public, however, is not the report itself (which would have a maximum circulation of 2,000 — members or subscribers, similar organisations, funders, academics, libraries) but through the mass media. Much effort is devoted to obtain good media coverage. But the criteria for evaluating just what is 'good' are vague: an op-ed piece in the *New York Times* will be counted as 'success'; the Morocco campaign will be judged to have 'bombed' in the media; the Tibet story did 'better than expected'. Few organisations have the time or resources to follow up the media or public impact of their reports. Often they keep no more than an old-fashioned album of press cuttings.

This gives no clue to how the media regime operates. How have journalists come to rely less on human rights organisations as primary sources (an Amnesty report is no longer the 'news' it was 20 years ago)? Why are some stories selected for coverage rather than others? What is the competition from other similar stories (aid, disaster, environment)? Why do some stories die a media death? And even if we knew how human rights issues are covered by the media, this does not tell us how the resulting stories or visual images are absorbed by audiences.

The reason why human rights organisations do not monitor this type of public communication is obvious: as soon as they end one campaign, they have to start working on another. Their timetables are structured from report to report. The fetishisation of the written text — scholarly, reliable and vivid as it might be — is a touching tribute to our faith in the power of knowledge. But it is a poor guide to public consciousness — let alone action.

Discussions about the impact of atrocity images or appeals from agencies like Amnesty, Action Aid, Oxfam or Greenpeace are less sophisticated than they look. Catch-phrases such as 'compassion fatigue', 'brutalisation', 'desensitisation', or 'donor burn-out' are neither proved nor provable. The hero of David Lodge's novel *Therapy*, going through another round of mailed charity appeals (with their 'smudgy b/w pictures' of starving black babies, kids in wheelchairs, stunned-looking refugees), reflects on compassion fatigue: '... the idea that we get so much human suffering thrust in our faces every day from the media that we've become sort of numbed, we've used up all our reserves of pity, anger, outrage and can only think of the pain in our own knee. I haven't got to that stage yet, not quite. But I know what they mean.'

We all do. But at another level, the concept is nonsense: you don't have compassion fatigue any more than you have love fatigue. Or else it is self-fulfilling: there is 'media fatigue' in the sense that journalists assume that their audiences don't want to hear more routine horror stories. But — as the recent events of Bosnia and Somalia as well as long-term trends in charity donations show — there are consistent reserves of compassion to be drawn on. And, much as those of us in the human rights movement would rather that this reserve be directed to people, the recent wave of support for animal rights hardly supports the thesis of compassion fatigue.

Human rights organisations and the media need far more solid findings about how their atrocity images are processed. I have been studying, for example, the extraordinary series of full-page newspaper advertisements produced by the Amnesty British section since 1990. These adverts are unmistakable and unforgettable, with their nagging, insistent quality, their emotional tone of 'quiet rage', their message of 'outrage into action' and their trade-mark sardonic trope: 'Death Camps. Cattle Trucks. Mass Graves. It's enough to make you write a letter of complaint.' 'Brazil has solved the problem of how to keep kids off the streets. Kill them.' 'You may well be tortured or killed when you get back to Sri Lanka. But that's no reason to feel persecuted.'

This type of message raises issues that, like the media imagery of atrocities, are largely unexamined. For example:

• *Shock tactics* What are the limits, pragmatic or ethical, to using gruesome images to attract cognitive and emotional attention? The 'severed head rule' dictates that certain images should not be used, because they are counter-productive, they 'go too far' in the exploitation of suffering, even for a good cause.

• *Simplification* Do certain effective communication techniques — like the personal case history — simplify the problem too much? Complex

STANLEY COHEN

Brazil has solved the problem of how to keep kids off the streets. Kill them.

Amnesty International newspaper campaign 1990s

political and legal issues may get lost when the issues are presented in an advert or direct mailing.

• *Emotional text* David Lodge's character talks of 'pity, anger, outrage'. Just what emotions are human rights appeals meant to arouse? The texts reveal three constellations: anger-rage-outrage, guilt-responsibility-shame and sympathy-empathy-identification.

• *Empowerment* Given the enormity of the problem — genocide, entrenched repression, ethnic cleansing, religious fundamentalism, collapse of government authority — how can ordinary people be given a sense that what they do can actually make a difference? This simple empowerment effect may override all the more complicated intellectual or emotional messages. I give my annual donation to Amnesty not because of the content of its direct mail letter (which I don't even read to the end), but because I believe that the organisation is doing a good job.

In the aid and development community, the equivalent debates have focused on the iconography of the starving African child. The debate is less advanced in the human rights field, nor are there images of equivalent power. Compared with both traditional charity appeals or the dramas of famine, aid and natural disasters, results are more difficult to convey. There are no obvious equivalents of the sick being healed, lives being saved or starving children being fed. Even the environmental movement, with its complicated long-term agenda, can point to immediate successes: dolphins saved, paper recycled, toxic waste dumps removed. The traditional empowerment claims made by human rights groups (a prisoner is released, torture or ill-treatment ended) are not only more difficult to prove, but are quite inappropriate to solving the mass atrocities of our time. Amnesty's narrative text of the prisoner of conscience being adopted and then released has no relevance to Bosnia, Burundi, Colombia or Liberia.

Ordinary people still, of course, respond to the human rights message — even on the less dramatic issues such as press freedom. And the otherwise remote successes of the UN enforcement apparatus (a new convention ratified, an investigation started) can be shown to have dramatic effects, for example, the beginnings of the War Crimes Tribunal to deal with violations in the former Yugoslavia. Nor, within the narrow 'conscience constituency' to which human rights groups appeal, is there necessarily competition between different causes: people build up a 'balanced portfolio' — Oxfam, Greenpeace and Amnesty.

But, even without explanations like 'compassion fatigue', it is self-evident that the sheer accretion of knowledge about violations (whether the drama of a political massacre in East Timor or the individual story of a house arrest in Burma) has not had the effect desired by those who collect all this information.

The political obstacles to the development of a mass-based human rights movement are, of course, beyond the reach of human rights organisations themselves. I am not suggesting that all we need is better reports. But if more factual information is not the issue — and the Rodney King effect showed that not even the clearest video images will produce the same response from everyone — then some serious thinking is required about the traditional modes of human rights reporting. Both the enterprise of 'witnessing the truth' and the efforts of organisations such as *Index on Censorship* to protect the 'right' to witness are informed by an Enlightenment faith in the power of knowledge. We need not accept the vanities and silliness of post-modernist theory to understand that the issue is not the abstract right to know, but what does it mean any more 'to know'. ❑

Stanley Cohen *is professor of criminology at the Hebrew University, Jerusalem. He has just been appointed Martin White professor of sociology at the London School of Economics*

This article is based on Denial and Acknowledgement: the Impact of Information about Human Rights Violations *by Stanley Cohen, available from the Centre for Human Rights, Faculty of Law, Hebrew University, Jerusalem 91905, Israel. US$10, including postage*

PADDY COULTER
Think pictures

THIS IS a plea to human rights organisations to get more serious about the mass media. Many organisations, as a matter of routine, devote impressive amounts of resources to researching, writing and publishing reports of human rights violations. The challenge is to commit commensurate levels of energy — and creativity — to communicating human rights issues through the media.

The starting-point for a non-elitist communications strategy has to be the candid acknowledgement that the mass media are the most important source of information for the public. Opinion surveys consistently show that the overwhelming majority of people derive their awareness of the wider world from television and, to a slightly lesser extent, radio and the press.

But — with a few very notable exceptions — most human rights reports are not targeted to audiences of millions but to tiny elite readerships or, at best, are put into the hands of that small select band of human rights specialists within the more upmarket print media. The impression given is that organisations are so drained by publication 'birth pains' that negligible time is left to plan innovative approaches to the mass media, and then more as an afterthought rather than an integral part of the project.

And yet innovation is precisely what is required as, at least in the industrialised countries of the North, trends in broadcasting are increasingly pointing in the wrong direction for human rights organisations. Intensified channel competition for audience ratings has led to increasing insularity and parochialism in programme-making, with a consequent reduction of in-depth coverage of international issues (always a relatively expensive business). In this Brave New Media world, human rights messages will be marginalised unless new initiatives are undertaken by human rights organisations to reach the man or woman in the street. And without evidence of a significant constituency of public support, politicians come under no real pressure to act on human rights violations, and, indeed, can dismiss human rights concerns as the preoccupation of a tiny elite circle of sympathisers and supporters.

So the single most important priority for each human rights organisation is to develop a serious, sophisticated and pro-active communications strategy of its own. To counter the prejudice generated in this article in favour of television, I would argue that newspapers have a dual importance within a communications strategy, not only for their own substantial readerships but also because they are regularly cannibalised by hard-pressed factual and features departments within television and radio; for some journalists there is a certain comfort to be derived from seeing a controversial human rights story in cold, hard, authoritative-seeming newsprint.

The key shift for human rights organisations is to generate stories (rather than issues), identified as priority targets for the media under its communications strategy. The mass media's appetite for human interest stories is enormous and within the human rights field there are large numbers of powerful and largely untold stories, but it may need a supportive outside journalist to advise here if the human rights organisation does not have the relevant expertise in-house. Using personalities to promote human rights messages is a well-established technique, but it does not have to mean rounding up the usual crowd of celebs: a wider canvas of less predictable folk could produce fresh media interest.

It should also be feasible to develop new ideas and material for radio and television programmes, and again if there is a dearth of good ideas internally there is no shortage of sympathetic independent producers hungry for broadcast commissions; (but do first examine their broadcasting pedigree in order to improve the odds of achieving results on air).

Above all, when planning major reports or campaigns, the most imaginative effort should go in at the outset to thinking up the visual imagery and sound bites which would most effectively accompany publication in print. Someone within the planning task force needs to be detailed to 'Think Pictures' and follow through in terms of video and stills photography.

A longer-term priority has to be the lobbying of the senior echelons of broadcasters and other media. This needs a different and sensitive approach if it is not to backfire. Some of this can best be tackled on an inter-agency basis, both for reasons of effectiveness and also to allay any suspicions of a corporate PR approach. The occasional round table between senior media executives and human rights organisations can be productive — as long as an effort is made to attract the real media 'movers and shakers' and not just the already converted.

But for all this to happen there has to be a real drive to confront the attitude prevalent in the more elitist reaches of the human rights community (not least among the funding organisations), that the so-called ephemeral media come a poor second to the printed report. ❏

Paddy Coulter *is director of the International Broadcasting Trust (IBT), a London-based, non-profit, independent TV production company specialising in programmes on development, environment and human rights issues for UK and international broadcast*

ROSALINDA GALANG

Keeping rights relevant

Human rights workers in the Philippines have been forced to adapt their methods to a changing world

IN ITS 20-YEAR history, the human rights movement in the Philippines has gone through two distinct phases. During its first decade of existence, under the Marcos dictatorship, our work concentrated on the massive violations of civil and political rights by a regime bent on stifling opposition to its rule. The work of getting the information about these abuses was both hazardous and painstaking. It required courage and considerable organisational skills. Ironically, the job of getting the information out to a wide audience was a relatively simple and straightforward one.

The martial law government sought to control all sources of information about the true state of affairs in the country and completely muzzled the mass media. Under these conditions, human rights information — as well as information about what was happening under martial law — became a precious commodity to both the local audience and the international community. The forms in which such information was communicated were often primitive: typed or mimeographed news-sheets, some clandestine, others under the protective cover of Church organisations. Copies were limited and distribution was laborious. But we had an eager and greatly supportive audience who spread the information in these news-sheets by word of mouth, by xerox or by whatever means was available to them.

We had no problems, either, getting the attention of and getting space in the international media. Moreover, a broad network of human rights and Philippine solidarity organisations abroad helped tremendously in relaying the truth about the Philippine human rights situation to a global audience.

During the 1980s, as the struggle against dictatorship gained headway, we were able to carve out space for alternative media. Non-governmental organisations, such as the Task Force Detainees of the Philippines, were able to come out with printed publications. Although circulation was still small — they were called the 'micromedia' — NGO publications worked in a similar fashion to the earlier news-sheets.

At this time, people's organisations in all the key sectors were rapidly expanding. They replicated the information and analysis contained in the micromedia to their grassroots constituencies — in public meetings, discussion groups or seminars, through manifestos, pamphlets, or educational materials. In this way, the micromedia, despite their limited resources, succeeded in reaching large audiences. And with the government still severely limiting access to information from official channels, NGOs were a major source of information for the newly emerging alternative press.

THE SITUATION changed greatly after the popular uprising of February 1986 which brought Marcos down and Corazon Aquino to power. By this time, there were more and bigger human rights organisations; the documentation of human rights information had become systematic and sophisticated; the alternative press had been transformed into the mainstream media. Yet it was during this period that the human rights movement experienced its greatest difficulty in communicating information about human rights violations and in reaching a broad audience. It even reached a point where the credibility of that information and of human rights organisations themselves were questioned.

With the free press back, NGO publications began to lose their appeal. It was not only that NGO writers and information people lacked the skills to compete with the professional media, or that they could not deliver the information as fast. They were also not able to take advantage of the democratic space which had opened to gain access to other sources of information than the traditional ones developed throughout the martial law years.

This diminished their ability to reflect swiftly changing realities. Human rights NGOs were turning out the same kind of information and analysis they used to provide during the dictatorship. They were not saying or offering anything new. Human rights violations were occurring in a totally different context, but the collection, packaging and analysis of human rights information were still being done within the old context.

As a result we found ourselves talking to a diminishing audience. The channels we had used so effectively in the past to disseminate information to a wider public had constricted. We were capable of producing glossy magazines, but we could not sell them. The publications only circulated within the international and local NGO community and even here, readership was static at best. As for our relations with the mass media, one journalist made this very harsh judgement: the NGOs — and that includes human rights NGOs — had become marginalised in the mass media. Just how far is indicated by the fact that the dailies did away with a regular human rights beat starting in 1991.

All of which makes me ask the question: does the low priority being given to human rights stories in the mass media simply reflect the public mood? With a government that has already signed practically all the international human rights treaties, and the statistical improvement in the human rights situation, do human rights still occupy a prominent place in the popular agenda? Is the constituency for human rights shrinking?

I believe there is still a broad constituency for human rights. That constituency may even be broader now because people have more access to information, are more knowledgeable about different aspects of our national life, have a richer body of experience and are more organised.

What prevents us from reaching and striking a chord in that broad constituency is not so much the methods that we use, but the kind of information we provide. Our systems of documentation are capturing only a small part of the whole human rights picture. We have developed powerful tools for gathering, recording and processing information, but what they are able to capture is a shrinking portion of Philippine reality. Reports of human rights violations arising out of a war in the remote parts of the countryside, or those arising out of repressive schemes by government are not only decreasing — they are no longer the whole or even the bigger picture of the human rights condition in our country.

The issues that we document and report on must be related to the bigger things happening in our society, such as the search for peaceful solutions to the armed conflicts that have torn our society apart, or the development path the present government is taking — or the things that have an impact on people's daily lives, like education, housing, jobs, health, the plight of our children, the devastation of the environment.

For the past three years, we at PhilRights, the information and

Manila, Philippines: community health project in city slum

research arm of the Philippine Alliance of Human Rights Advocates, have been going in this direction. We are not only improving and broadening our coverage of civil and political rights, we are also extending our coverage to economic, social and cultural rights.

We have seen some promising results. Two years ago, when we put our bi-annual publication, *Human Rights Forum*, in the bookstores, very few believed that it would sell. Yet from practically no sales at all we are now able to sell 35-40 per cent of the total number of copies, and about 65 per cent of these are through commercial outlets. Only 30 per cent are accounted for by subscribers in the NGO community, both local and international. Our best-selling issue thus far is the one on development and human rights.

There is another hopeful sign: increasing and lively participation in recent human rights forums with topics like 'development aggression' and the 'universality of human rights'. We are finding the human rights issues that are relevant and significant to a broad audience. And having found them, we shall certainly find the most effective means of reaching and communicating with that audience. ❑

Rosalinda Galang is editor of Human Rights Forum *for the Philippine Human Rights Information Center. She has been a public information officer for Task Force Detainees of the Philippines. She also has been a journalist with several publications and a research associate for the Population Center Foundation.*

CAROLINE MOOREHEAD & URSULA OWEN

Massacre in Rwanda, 1994

Time to think again

Why is an ever growing group of articulate, dedicated people achieving so little?

On 6 April 1994 the plane carrying President Habyarimana of Rwanda and President Ntaryamira of neighbouring Burundi was hit by a rocket. They were both killed. Within hours, Rwanda's Hutu majority had started on a massacre of the Tutsi minority; by the end of the summer nearly a million people were dead. The reaction, throughout the world, was one of horror. But it was also one of surprise. Where had this genocide suddenly sprung from? The answer was that it had not sprung from anywhere. People knew it was coming. It had been prepared, orchestrated, long in advance. Well before the day came when it took just one incident to unleash a systematic programme of murder, warnings had been sounded, reports written, governments lobbied. But warnings are not news; and warnings do not make the kind of stories that appeal to news editors.

No-one any longer denies — not even the governments who carry them out — that human rights violations are occurring today at an overwhelming rate. In the quest for foreign markets, many professed commitments to human rights are being forgotten. 'Commercial Diplomacy' is the phrase of the moment. Nor does anyone deny that knowledge of these events has never been greater. No other generation has ever known so much, so quickly, so graphically. There have never been so many reports, on every violation and every country, so well written and so accessibly presented. Human rights organisations do invaluable work — they are needed as never before. Why, then, as a group of human rights experts asked themselves at a seminar in Oxford in July, is all this knowledge having so very little effect? Has more knowledge become less action? Are the images and reports too disturbing? Or not disturbing enough? Are human rights groups concentrating too much on producing the report and not enough on getting the report's views out to a wider world, as Stan Cohen suggests in his impressive report *Denial and Acknowledgement: The Impact of Information about Human Rights Information*, around which the Oxford conference was centred. Why, in short, is an ever-growing group of articulate and professional people, trying to stop human rights violations, apparently achieving so very little?

There are, of course, no easy answers. When Amnesty International was founded 35 years ago the world it confronted, if no nicer, was more straightforward. Amnesty hit on what is probably the most inspired concept ever dreamt up by the human rights world: the fact that the public needs to identify with individuals and their personal stories, and that if allowed to do so, will go to great lengths to work on their behalf. That formula reached its high point sometime in the 1980s, when crusades on behalf of individual victims were drawing in supporters all over the western world. When Safia Hashi Madar, a 26-year-old biochemist from Somalia, was suddenly released from detention and torture in the spring of 1989, and the prison governor asked her angrily how it was that she had so many friends in the West who had written in their hundreds begging for her release, it was a personal victory for everyone who had ever worked to secure a prisoner's freedom.

Thirty-five years on, it is all rather different. The end of the Cold War has changed some of the rules and Amnesty International has been joined by hundreds, perhaps even thousands, of other organisations, both national and international many of whom are turning out their own excellent reports. Whereas, in the late 1970s, there used to be a single report, on, say, atrocities by security forces in Kashmir, there are now five, making them in some curious way easier to disregard by foreign editors pressed for space. The publication of a new report, on torture, genocide or street children, in a world bombarded by atrocities, is no longer in itself an event. Television has brought the slaughter straight on to our screens. The question, today, is no longer whether it is possible to get hold of a photograph of a massacre of Peruvian villagers by the security forces, but whether it is right to show their severed heads stretching in a long line along a dusty street.

In any other field of human endeavour, such meagre results as are being seen today for such an enormous quantum of effort would not be tolerated

As the fiftieth anniversary of the UN's Declaration of Human Rights approaches, there is a sense of crisis in the human rights world, for, as with humanitarian aid, no-one, any longer, knows where to go next.

In Oxford, in July, there were suggestions. The human rights reports are now excellent and do not, for the most part, need improving on. What does need improving, is not just the dissemination of the reports but their hit rate. They have to be made to matter again, to touch people's feelings, in the way they did 35 years ago. On their own, they have lost their resonance. Human rights workers have to attend as much to the report reaching the media, in ways that the media find acceptable and usable, as they do to accuracy in the report. And the media, as we know, does not always make it easy. Paddy Coulter (page 46) puts down a few ground rules, which are rarely followed by human rights groups. Funders, who make little provision for the distribution and impact of the work they pay for, must be urged to budget for the last, most often neglected, stage, that of making them count. It is this disparity between what is known and reported, and how it is communicated and acted upon, which has become so glaring.

The relationship between international and national NGOs also remains uneasy, with smaller grassroots organisations questioning what they see as an attitude of superiority and arrogance on the part of the western, long-established bodies, an unwillingness to listen to those fighting it out on the ground. A sense of co-operation between organisations, even where resources are limited, may now be one of the only ways to secure the necessary attention — without which there is a danger that the entire exercise will gradually become irrelevant. And yet human rights organisations admit that they sometimes resist co-operation — often over anxieties about funding, sometimes because of anxieties about their profile. Specialisations among the organisations, a sharing of strategies and joint campaigns, competition over funding, the proliferation of seemingly similar bodies, and how to bring in the young — these are all questions that may need reopening.

And the time may have come to look at other models, more populist ones perhaps, outside the human rights world. Electoral campaigns, for example, have a far cruder measure of success; they set out to win. Their methods are very different from the tentative steps taken by human rights organisations. They send out mailing, they hustle door to door, they make speeches, they try to persuade. Could this injection of immediacy return to the public a feeling it clearly had when so many people rushed to join Amnesty International in the 1970s, an understanding that human rights are not something that happen to other people, a long way away, but concern everyone, everywhere? The assumption, since the UN signed its historic declaration half a century ago, has been that if people knew enough, if the quality of research and reporting was good enough, if the information reached the right people, then action was bound to follow. That is turning out to be false. In any other field of human endeavour, such meagre results as are being seen today for such an enormous quantum of effort would not be tolerated. Perhaps the moment has now come for the human rights movement to meet, to talk, to question and to think again. ❏

Caroline Moorehead is a writer and broadcaster specialising in human rights

BOOKS FOR THE BURNING — OR BUYING
The *Index* Auction of Banned Books

DRAWN by one of the most intriguing and extraordinary collections of books ever assembled for sale, dealers, collectors and *Index* supporters converged on the Middle Temple in London on 6 November 1995.

Authors on sale included George Orwell, Erica Jong, Judy Blume, Boris Pasternak, Ken Follett, Tom Paine, Percy Bysshe Shelley and a unique collection of *samizdat* from the former Czechoslovakia. The writers were united by one thing — they were all the victims of censorship.

Our right to choose freely what we read has been fought out in the courts over the centuries, so it was fitting that the *Index* Auction of Banned Books should be set in one of London's ancient Inns of Court.

The hammer was brought down by Lord Hindlip of Christie's, assisted by guest

From Lady Chatterley to Satanic Verses: Salman Rushdie at the auction: kindred spirits

auctioneers Anna Ford and Frank Delaney. Bidding was brisk, and the stars of the evening turned out to be James Joyce and D H Lawrence. A rare catalogue, *The paintings of D H Lawrence*, made £1,550, closely followed by a first edition of *Lady Chatterley's Lover* at £1,500.

The evening raised £20,000 for *Index* and Writers and Scholars Educational Trust.

Joe Hipgrave

LETTER FROM LAGOS

ADEWALE MAJA-PEARCE

The murderers in our midst

'I have devoted all my intellectual and material resources — my very life — to a cause in which I have total belief, and from which I cannot be blackmailed or intimidated.'
Ken Saro-Wiwa, 1941-1995

It seems that the coffins were ordered and the hangman and his three assistants flown in from the far north two days before the death sentences on Ken Saro-Wiwa and his fellow Ogoni activists Barinem Kiobel, John Kpunien, Baribo Bero, Saturday Dobue, Felix Nwate, Monday Eawo, Daniel Kbakoo and Paul Levura — were confirmed by General Sani Abacha's Provisional Ruling Council, the government in this case being both the prosecutor and the final court of appeal. Two days later, at 9.30 in the morning, Friday 10 November, the Ogoni Nine, as they had come to be called, were taken under armed escort from the military camp they had been held in for the last 18 months to the regular prison in Port Harcourt. They were ushered into a large cell which was empty save for a long wooden bench. They hadn't yet been told about their impending executions. Along with the rest of the world, they never imagined that the government would kill them with such indecent haste, and on the very day the Commonwealth was meeting in Auckland. A few minutes later, an inner door opened and Ken was beckoned inside. He was immediately confronted with a priest, who proceeded to give him the last rites. Then he was asked whether he had a final request. He asked to see his wife, Hauwa. This was denied. He asked to see his 91-

year-old father to give him his pipe and his wallet. This, too, was denied. Then the sheriff read out the death-warrant, then a black hood was placed over his head, then the noose was tightened around his neck.

And then the story becomes bizarre. 'It failed,' a 'source' was reported to have said. 'Try and try the hangman did, but it simply failed to work.' Apparently, the lever refused to release the trapdoor, although it had been successfully tested on a dummy that very morning — for the last four days in fact. They decided to try it out on one of the others so they led Ken back to the cell and brought out Kpunien. This time it worked. They brought Ken back again and again it failed. It wasn't until the fourth attempt that the lever slammed home, as it were, whereupon the military administrator of Rivers State, Lt-Col Dauda Komo, the same man who had publicly pronounced Ken and the others guilty even before the start of what passed for a trial; and who, now, at Ken's last hour, insisted on being personally present at the sickening spectacle, rushed down the makeshift scaffolding to make sure that he was well and truly in possession of his corpse. According to another source, the hangings were videotaped, presumably for the entertainment of those who think nothing of human life; at any rate, they wanted Ken dead, and they killed him, as he knew they would. 'My murder is being officially planned and executed,' he wrote within weeks of his incarceration, but then Ken was in a better position than most of us to know the manner of people who would rule this country by force of arms alone, which is why they talk endlessly of ensuring the nation's peace and stability, being themselves the greatest threat to our security.

'I do not blame Mandela because, having spent 27 years in detention, he has lost touch with the global socio-political trend.' This was Abacha's response to the South African president's call for Nigeria's suspension from the Commonwealth. Honourable ministers promptly followed suit. One of them, Iyorwuese Hagher, the minister of state for power and steel, argued at length in a newspaper article that Mandela's action had shown him to be the house nigger of 'the new slave masters', ie the old slave masters, including 'the South African apartheid machinery'. Witness, for instance, how easily they had wrested him 'from the control of the black militancy of Winnie' in order that he might better 'perform the dance of the stooges'. Another minister, this time for agriculture, Professor Admola Adeshina, went even further. 'How can somebody spend 27 years in prison and still be sane?' he queried, and then, working

LETTER FROM LAGOS

African Writers Conference, Potsdam 1992: Left to right: Andrew Graham-Yooll, Adewale Maja Pearce, Jack Mapanje, Biyi Bandele-Thomas, Ken Saro-Wiwa

himself up to a state of complete hysteria, pronounced Mandela 'a white man in black skin and no white man likes the blacks'. Indeed: Mandela's own description of the regime as 'illegitimate, barbaric and arrogant' missed out only the word stupid.

And then it turned out that it was all a dreadful misunderstanding: that Abacha hadn't meant anything derogatory in his reference to Mandela's long years of imprisonment; and that what actually happened was that, 'the statement...was not in fact his exact words', but that 'the wordings (sic!) were amplifications by the press of what the head of state said when he met with traditional rulers and leaders of thought.' This was from Abacha's chief press secretary, David Attah, indulging the national pastime, which is that it's always somebody else's fault in Nigeria, especially when the latest victim of the country's terminally irresponsible press also happens to be 'a mature, calculating and self-respecting statesman'. As Ken said of the judge who sentenced him to be hanged by the neck until he was dead: 'The doctor told him I was having psychiatric problems. I said yes, I must be, because all I can see on the benches are kangaroos.'

THE MILITARY tribunal that killed Ken Saro-Wiwa and the eight others was established by the Civil Disturbances (Special Tribunal) Decree. It has been on the statute books since 1987, long before anyone had heard about the rape of oil-rich Ogoniland in southern Nigeria. But military tribunals have been killing civilians for 20 years, and it is only now that the civilian judges who confer on them a spurious legitimacy are beginning to wonder whether the rule of law isn't better served by the regular courts. 'On trial also is the Nigerian nation,' Ken wrote in a statement he was prevented from reading at his trial. 'The military do not act alone. They are supported by a gaggle of politicians, lawyers, judges, academics and businessmen, all of them hiding under the claim that they are only doing their duty.' Even as the hangman from the north and his three assistants were testing out their dummy, but before Abacha had confirmed the sentences, the chief judge of the Federal High Court in Lagos, Justice Babatunde Belgore, airily brushed aside a petition for a stay of execution pending a decision from the African Commission on Human and Peoples' Rights, to which Nigeria is a signatory. I spoke to the lawyer who pleaded with him for over an hour in his chambers to render a quick decision. 'My friend, who told you they will kill them,' he said, and set a date for a hearing which turned out to be exactly one week too late.

'I have no doubt at all about the ultimate success of my cause, no matter the trials and tribulations which I and those who believe with me may encounter on our journey,' Ken wrote in his final message, but there's little chance that the time will be soon, with or without Abacha at the helm. The level of 'debate' exercised by the Honourable Ministers — civilians both, and one a professor — in response to the country's most serious external crisis since the civil war in the '60s (and perhaps not even then) is hardly bettered among our 'born again' politicians now seeking — at this late hour! — to bury their differences in the defence of democracy, etc etc. Students, journalists and human rights activists are, of course, making all the usual noises, and, as usual, paying the price for it, but there's nothing new in that. They've always done their job. Would that others did theirs, if only for the sake of the memory of a man who gave his 'very life' that we might rid ourselves of the murderers in our midst. ❑

Image makers

I BUY three newspapers a day. My monthly bill comes to about N3,000, US$36. This is more than most people earn for food and rent and school fees, but it's the only way to get a reasonably balanced picture of what's going on in the country. Despite the existence of four independent television stations and an independent radio station, all of which steer clear of politics, the electronic media are still effectively in the hands of the government.

Now, of course, with Nigeria suspended from the Commonwealth and the international community breathing fire and brimstone, General Sani Abacha's beleaguered administration is pulling out all the stops. Rumour has it that the government is planning to spend US$50 million for advertisements in selected newspapers in Europe, the USA and South Africa (Zimbabwe has already been 'done') on what it calls 'an image-laundering exercise'.

But not all our money — or what's left of it — will be leaving our shores. The government has also announced plans to upgrade the facilities of the Nigerian Television Authority, aka Nigerian Trash Amplifier, in order to convince the home market that all the lies directed at our fatherland are really part of a sinister design to re-colonise the country in order to deny it a permanent seat on the UN Security Council.

It's not altogether clear what the government itself believes, if only because its own sources of information are often unreliable. The government-owned *Daily Times*, for instance, whose facilities are also being beefed up the better to counter 'the country's negative image in the western press', led with a report of Japan's support for Nigeria at this trying time in the nation's march towards greatness. Its source was the Japanese chargé d'affaires who himself denied the story as 'a complete misrepresentation', 'a complete fabrication', in an interview with *The Punch*, an independent daily.

The Punch, along with *The Guardian,* has just been unbanned after more than a year in the 'cooler'. And since the government is in no position to take on yet more enemies, especially with all this talk about oil sanctions and freezing the assets of military officers, better simply to ban the popular early morning press review programmes on all 31 state-owned radio stations. Some radio stations, it appears, are resisting the directive. One of them, Ogun State Broadcasting Corporation, 'still reviews newspapers every morning, but in the past few days, words such as 'sanctions', 'Ogoni', 'Commonwealth' and 'Mandela' can no longer be heard. **AMP**

VLADIMIR VOINOVICH

The gang of four

America? Land of immigrants? Not any more sir

I FILLED the questionnaire in on the flight. Last name; First name; Passport number; Purpose of your visit to the United States: a) pleasure; b) business. As always, I ringed 'pleasure', since what I do in the United States is not my understanding of 'business'.

The Immigration Service queue was quite short. I handed the officer my German passport, every aspect of my appearance proclaiming the honourableness of my intentions towards his country.

My passport may look a bit odd, but there is nothing sinister about it. It is issued by a country with which the United States of America has an agreement permitting return journeys without a visa. In any case, I have B-1 and B-2 visas. There is even a J-1 visa which, although expired, would tell an alert official that the bearer had recently lived for a year in Washington, DC in full accordance with the law and occupying himself respectably at the Woodrow Wilson International Center for Scholars. Taken together these visas testify unambiguously to my having entered the USA on numerous occasions, spending from a few days to more than a year there, and never having been detected undertaking activities prejudicial to the state or attempting to overstay my welcome.

So I was not anticipating any trouble as I presented my German passport to an official with the cropped head of a greying mugger who, at two o'clock in the afternoon, might not be expected to have worked himself into a state of stupefaction or, at roughly 50 years of age, to have abandoned all hope of further advancement in his career. For some reason he took an instant dislike to me and said without more ado that my passport and visa cut no ice whatsoever, and he would have to give further consideration as to whether I was worthy to set foot on the territory of the United States. Then he asked to see my return ticket. I didn't have it.

On 20 March 1993 I had flown to America from Munich for a seven-week stay, with a return ticket for 8 May. I had flown to Paris on 20 April and back on 24 April on a ticket bought in Baltimore and exchanged for boarding cards between Baltimore and Paris. As a result all I had to show on my return were a couple of counterfoils and a USAir folder. I tried to explain the situation. He did not understand. I suggested the problem might lie with my English, but he rejected this as a cheap ploy: 'No, your English is good enough.' For some reason Mugger got more and more worked up, announced that he did not believe a word I was saying, wrote some comments on a slip of paper and sent me off to see the supervisor, a young black woman whose generous proportions convicted her of frequent and immoderate consumption of popcorn. Her colleagues called her Judy.

As I made my way towards Judy I remembered that reply in my questionnaire and began worrying that my six-week college seminar probably did count as 'business'.

Judy flashed me a broad and engaging smile and asked who was supporting me financially in the United States. I replied that I did not need anybody to support me because I had only two weeks of my stay remaining and would be quite capable of feeding myself unaided. To prove the point and put an end to all this questioning I decided to show her my credit cards. I opened my wallet but... Oh horror! Would that a mere serpent had darted from it! Out fell my temporary accreditation as a visiting professor of the college where I conducted my humble seminar. I was as flummoxed as if the card had been a forgery and tried to push it into another pocket of the wallet, but Judy was too vigilant. She whipped the card out, examined it, frowned deeply, and from that moment it was downhill all the way.

Having studied my card Judy informed me without more ado that I was a liar, a dishonest person, that they had not finished with me, and that meanwhile I could sit and wait until all the passengers for Baltimore had been through passport control.

I had had a sleepless night and was exhausted by my journey. I felt very low and anxious, knowing that the two women waiting for me at the exit from the customs and immigration hall must be worrying about my long non-appearance. I was worried about it myself. I asked Judy to tell the people meeting me (without saying who they were) that I had arrived.

Her reply was unexpected: 'I am not concerned with who is waiting for you and where.' Some time later I repeated the request and was informed that she occupied an official position and did not run unofficial errands for persons as dishonest as I. I expressed surprise and asked whether, while occupying an official position, she was not also a human being. This upset her and she assured me, 'I am a good person', and one, moreover, who never lied and had no intention of responding to requests from people who did. After this she asked me to think and say honestly who I was. I replied that I honestly thought myself to be an elderly male German citizen and Russian writer, quite well-known, even to a few people in the United States. Judy instantly recognised this as a blatant lie. I repeated the information to Mr Mugger. At this he cackled exactly like a Russian hen, 'Ha-kda-kda-kda!', and gestured dismissively, not wishing even to listen to such manifest nonsense. He then produced a roast chicken from under the counter and began tearing it apart with such frenzy that Sigmund Freud, had he been present, would have been agog to know what passions were seething at that moment in his subconscious. I recognised the futility of talking to Mugger and again turned to Judy. I told her that, while not wishing to exploit the fact, I had had a heart operation, and by refusing to tell the people meeting me where I was they were making a heart patient, who was already extremely tired, very nervous and endangering his health. Not having any documents about the operation, I bared my sunken chest and displayed its scar. I do not

essential READING

—**INDEX** subscriber

Lionel Blue
Joseph Brodsky
A S Byatt
Beatrix Campbell
Noam Chomsky
Alan Clark
Emma Donoghue
Ariel Dorfman
Ronald Dworkin
Umberto Eco
James Fenton
Paul Foot
Zufar Gareev
Timothy Garton Ash
Martha Gellhorn
Nadine Gordimer
Gunter Grass
Vaclav Havel
Christopher Hitchens
June Jordan
Ryszard Kapuscinski
Yasar Kemal
Helena Kennedy
Ivan Klima
Doris Lessing
Mario Vargas Llosa
Naguib Mahfouz
Alberto Manguel
Arthur Miller
Caroline Moorehead
Aryeh Neier
Harold Pinter
Salman Rushdie
Edward Said
Posy Simmonds
John Simpson
Alexander Solzhenitsyn
Wole Soyinka
Stephen Spender
Tatyana Tolstaya
Alex de Waal
Edmund White
Vladimir Zhirinovsky

United Kingdom & Overseas (excluding USA & Canada)

1 year—6 issues	**UK:**	£36	**Overseas:**	£42	**Students:** £25
2 years—12 issues		£65		£77	
3 years—18 issues		£95		£113	

Name

Address

Postcode   B6A1

£ _____ total. ❏ Cheque (£) ❏ Visa/Mastercard ❏ Am Ex ❏ Diners Club

Card No.

Expiry   Signature

❏ I would also like to send **INDEX** to a reader in the developing world—just £23. These sponsored subscriptions promote free speech around the world for only the cost of printing and postage.

❏ I do not wish to receive mail from other companies.

INDEX, Freepost, 33 Islington High Street, London N1 9BR **Telephone:** 0171 278 2313

United States and Canada

1 year—6 issues	**US:**	$50	**Students:**	$35
2 years—12 issues		$92		
3 years—18 issues		$130		

Name

Address

Postcode   B6B1

$ _____ total.
❏ Check (US$) ❏ Visa/Mastercard ❏ Am Ex ❏ Diners Club

Card No.

Expiry   Signature

❏ I would also like to send **INDEX** to a reader in the developing world—just £23. These sponsored subscriptions promote free speech around the world for only the cost of printing and postage.

❏ I do not wish to receive mail from other companies.

INDEX ON CENSORSHIP

33 Islington High Street, London N1 9LH England   **Facsimile:** 0171 278 2313

BUSINESS REPLY SERVICE
Licence No. LON 6323

INDEX ON CENSORSHIP
33 Islington High Street
London N1 9BR
United Kingdom

NO POSTAGE
NECESSARY
IF MAILED
IN THE
UNITED STATES

BUSINESS REPLY MAIL
FIRST CLASS PERMIT NO.7796 NEW YORK, NY

Postage will be paid by addressee.

INDEX ON CENSORSHIP
215 Park Avenue South
11th Floor
New York, NY 10211-0997

remember what she replied, but Mr Mugger, gnawing at a chicken bone, said, 'Don't try to intimidate us.' I could find no immediate riposte to this, although my chest scar does not look all that alarming and could hardly have been used to intimidate so intrepid a militant as him. (I do have a potentially more intimidating scar on my leg, but practical considerations precluded rolling my trouser leg up high enough to expose that to him.) Mugger continued gnawing his bone with an expression of immense satisfaction, perhaps derived less from the chicken he was demolishing than from an ambitious surmise that he had caught a spy or major international criminal like General Noriega of Panama, and stood to be made a general himself.

A THIRD, female, person of southeast-Asian origin now appeared on the scene who resembled Mao Tse-tung's widow, famed leader of the Gang of Four. This new arrival made no move to introduce herself, and in papers subsequently issued to me identified herself only as No. 33.

With the arrival of No. 33 the situation sharply deteriorated. I felt I had landed in a lunatic asylum, or returned to that distant past when my personal file was being reviewed in the Union of Writers of the USSR. 'You are not telling the truth. That is a lie. Where is your return ticket?' I would start trying to explain: 'My return ticket...' 'Stop! We do the talking... What is the purpose of your visit to the United States?' 'I have come to the United States...' 'No, we do not believe you. Who are you really? Answer, answer at once. Who are you and what is the purpose of your visit to the United States?'

By now I no longer had any grounds, motive, or possibility of concealing anything whatsoever, but I had only to open my mouth to be instantly silenced.

I explained that I conducted a seminar or conversation class (I was not sure which better fitted my visa) in a college twice a week for students who were studying Russian on the basis of my books.

My interrogators did not believe me and, looking daggers, repeated again and again, 'Who are you? What is the purpose..?', until I began to feel that perhaps I really had come to the United States with some sinister intention which I had concealed from myself. I tried, against all the odds, to communicate: 'I repeat, I have come...' 'Don't repeat anything. You don't want to tell us the truth. We are putting you on the next plane back to London.' 'Fine, send me back to London, but before you do can you tell the people waiting to meet me where I am?' 'Whatever next! We are not here to run around looking for your friends.' 'But my friend works at that college. She can confirm what I am saying.' 'We do not need anyone to confirm anything. We are sending you back to London.'

From time to time my investigators would confer hurriedly and in considerable excitement among themselves. I gathered they had rejected their initial supposition that I was an aspiring illegal immigrant in favour of something more appropriate to a detective novel, in which connection the initials of the FBI were ominously mentioned several times. I do not know whether they did phone the FBI (I rather think they did), but I do know they did not phone the college, where they could have expected to obtain all the information they needed in one go and where they might have discovered that the criminal they were dealing with was at least not as hardened as they imagined.

I suspect that they did in fact realise at some point that I was who I said I was, but that the further they blundered down their dead end the more vicious they became. No. 33 kept running off somewhere and returning with a new question to throw at me in a tempestuous manner, lapsing at times into total absurdity. 'How often?' 'How often what?' 'I am asking you, how often?' 'And I am asking you, how often what?' 'You don't ask the questions round here, and I am asking you, how often?' 'How often? I don't understand.' 'You understand perfectly well. How often?'

'How often what?' I yelled. 'How often do I clean my teeth or wash my feet, or what?'

It transpired that she wanted to know how often I gave lectures. I explained I didn't give lectures, I gave conversation classes… Without letting me finish she rushed over to my open travel bag which contained three versions of a screenplay in Russian and one in Czech. 'What are these? Are these your lectures?'

Oh, Lord! How could I bring it home to this stupid, ignorant woman that no lecture in the world could possibly look like this screenplay? I kept saying to her, 'Phone the college or ask the people waiting to meet me and everything will become clear to you.'

'No, we do not need to talk to anyone, we do not need to phone, we shall get everything out of you.' I did, nevertheless, start trying to write down the college's telephone number, but failed. Never in my life had my hands been shaking as they were there. What sort of brutish, heartless

USA: GANG OF FOUR

people were these? I had already told them about my heart operation, which I had not by any means undergone in order to bamboozle the US Immigration Service.

At some point I told No. 33 I would answer no further questions from her and demanded to see a doctor, the people waiting for me, a lawyer, a German consular official, a Russian consular official, or a representative of the US State Department. Mr Mugger guffawed and said such demands could be made only by American citizens. I said I would answer no further questions anyway. 'Why?' No. 33 shouted. 'Because you do not listen and just try to use my answers against me.'

In the course of the interrogation I had placed all the papers I had on me on the table. No. 33 now decided to make copies of these. They suddenly discovered that that damned card from the college which had started this whole thing was missing, and began looking under all the papers scattered over the table. No card.

All three then turned towards me, boring into me with their eyes, suspecting that I had filched (my own) card from them. I voluntarily turned out my jacket and trouser pockets and laid everything, down to my loose change, on the table.

No. 33 drummed her fingers on the table as if playing the piano and a fourth person instantly materialised, a massive young man with a bull neck, bulging biceps, and a small pistol on his belt. I tried to read the name on his identity badge: Mincara, I think it was.

I don't remember how I came to find myself face to face with a glass partition, my hands pressed against it and my legs spread wide apart.

This was something I had seen in films about gangsters, but had never imagined it would happen to me.

There I stood, an old, grey-haired man, squashed against their partition with my hands up. Other passengers were walking by, among them perhaps some of my compatriots, some, indeed, who might recognise me. My vanity was stung by what these people passing by might think of me (and some did stop with their suitcases and stare at what was going on with great and understandable curiosity).

Mincara prodded me in the back and shouted something, calling me 'Sir' in the process. I felt a pain in my chest, and again began mumbling something pathetic about my heart operation. 'Sir,' Mincara assured me, palpating my right arm from hand to armpit, 'Sir,' he said, moving to the left arm, 'This will not take long, sir. I am treating you carefully, sir. Do not arch your back, sir. Do not turn your head round, sir, do not turn your head round, sir, I am telling you not to turn your head round.' Sir, however, insisted on turning his head round, reluctant to lose sight of the bag with his papers. Sir had not the least doubt that what was sought would shortly be found, and not in his possession. Moreover, sir was far from persuaded that these worthy citizens would not then plant the missing card in his bag.

Having fingered me from top to toe, Mincara ordered me to remove my socks and shoes.

'And now, sir,' he said, 'hand me your shoes!'

To tell the truth, I was in a sweat. I had no intention of bending down to undo my shoes, but what would be the outcome if they started forcing me down? I preferred not to think about it.

'Sir,' I heard, as if from far away, 'I repeat, hand...'

Suddenly a hysterical shriek:

'Stop, Stop, Stop!'

No. 33 had finally found my card from the college.

There followed a re-enactment of the dumb scene at the end of *The Government Inspector*. No. 33 scrutinised the card, apparently in the hope that it would somehow vaporise. Judy beat a retreat, while Mr Mugger froze with an expression on his face which suggested he had swallowed a live chicken, feathers and all.

Then I did start shouting and demanding an apology.

Mincara was in evident perplexity. He suddenly took to his heels, shouting to me as he disappeared, 'Have a nice day!' I shouted unkindly after him that I wished him nothing of the sort in return. Mugger (whose name I later discovered with great difficulty was Broker), also absented himself as a matter of urgency and was not seen again. I addressed myself to No. 33 and asked whether she might wish to apologise. Judy shouted in her place, 'I apologise! I apologise!' and added aggressively, as if she had just caught me up to something else: 'Do you accept it?'

No. 33 meanwhile took herself off. It turned out she had finally gone to see who had come to meet me. She returned with my wife's passport and again started shouting at me for not telling her my wife was waiting for me. She was evidently trying to find some justification at least for the behaviour of the gang of four whose leader she was, but it was inexcusable. I could feel the atmosphere gradually returning to normal. No. 33 came back to the question of my papers and, when I again mentioned the screenplay, she smiled ingratiatingly and even gave a little skip like a schoolgirl confronted by Arnold Schwarzenegger: 'Ah, you are making a movie?' However, she immediately remembered herself and the glint of hostility came back into her eyes. She sternly informed me that I was being refused entry to the United States, that my documents were being confiscated and would be returned to me on Wednesday 28 April, when the question of my deportation would be decided.

That Wednesday I presented myself to the same officials, but this time accompanied by a lawyer, and bearing two of my books and a recent article from the *Washington Post*, which proved useful. When we arrived the door was opened by No. 33, by now looking very down in the dumps. She had had various telephone calls enquiring about the details of the incident, and advising her that I would be demanding a written apology. I heard later that she was very anxious. 'Hi,' I said, but got no reply. Judy went past looking as though she had just buried a much-loved cat.

Mr Mincara asked the questions, but this time did not put me up against the wall and even seemed a little contrite. When he dropped a piece of paper he bent down and picked it up himself, which, given his great size, was not that simple an act to accomplish. This time only a few minutes were needed for him to elicit the truth. While he was thus employed Judy Thomas (the lawyer told me her surname) approached and started mumbling in self-justification. In recognition of her earlier apology, albeit not written, I cut her short and said I had nothing against her personally. 'I also have nothing against you, sir!' she squeaked joyfully. Hearing this No. 33 also came running (she proved to have a name after all: Shi Bao), evidently hoping for an indulgence too, but as she immediately started berating me again, she got the cold shoulder.

In possession of my passport once again, I went out into America thinking how fortunate I was to have met these people. They had shown me how easily any trip can turn from pleasure into business for a satirist. Not very profitable business, however, as I could find no American magazine or newspaper to print my story. The editor of one well-known newspaper took it, but returned it on the grounds that his readers would not understand why I had incorrectly stated the purpose of my visit to America. But I went to America entirely legally, violating no laws and harbouring no criminal intentions. The inconsistency in my questionnaire (even if they did find my explanations unconvincing) hardly gave the official muggers of Baltimore Airport the right to try to prise the truth out of me in such a churlish fashion.

PS. America is currently experiencing difficulties said to have many causes, including an influx of illegal immigrants. America is a country of immigrants and in earlier years encouraged new arrivals. Now, by force of circumstance, she is fighting them in spite of her traditions and ideals. In the forefront of this struggle stand the officials of the US Immigration Service, endowed with limitless powers but not burdened with much responsibility for their actions. Among these officials you do, no doubt, encounter all sorts, but the proportion of sedulous nerds and outright bastards is manifestly way over the norm, a proposition readily confirmed by a visit to the consular section of the US embassy in Moscow.

One of my friends was fortunate enough to be born in the United States of America, and unfortunate enough to be transported in infancy to the Union of Soviet Socialist Republics. If previously he had, for obvious reasons, kept quiet about his origins, now in his 50s he recently produced his American credentials and received the passport of a citizen of the USA. He emigrated to America, found a job, and wrote to his wife to join him. She is being denied entry until she can prove that her marriage to this American citizen is not fictitious and undertakes not to remain permanently in the United States. The officials are unmoved by her 30 years of marriage, since no regulation stipulates what period of marriage can be taken as a proof of its non-fictitiousness. As for undertaking not to remain there permanently, the situation is even more parlous, since she hopes that when she rejoins her husband they will remain together for the rest of their lives; something she has now been fruitlessly trying to achieve for many months. ❑

Vladimir Voinovich, *a dissident and writer, was excluded from the Writers' Union in 1974 and left the USSR in 1980. He now divides his time between Munich and Moscow.* © *Translated by Arch Tate*

MUMIA ABU-JAMAL

The power of money

EQUAL JUSTICE

THE QUICK VERDICT in the O J Simpson trial has ripped the covers off the naked flesh of race in the US criminal justice system, and in society as well.

Factions, sectors and divisions have erupted in all areas of American life, with fury, born of fear.

The Simpson verdict did not create these factions, sectors and/or divisions — it merely highlighted those already in existence.

Depending on the perspective of who speaks, the O J verdict is seen as either: a rare victory for Blacks in court; or an injustice.

Why is it so rarely seen that the O J verdict is only truly possible to those who possess enormous wealth?

Why was his acquittal a 'Black victory' if it literally cost almost US$4 million?

How many Blacks can raise such an outlandish sum, even if their lives and liberties are at stake?

Make no mistake, dear readers, O J Simpson is a free man today because he was a multi-millionaire whose wealth was utilised to secure his freedom. Like Claus von Bülow, the Rhode Island socialite who spent an estimated US$ one million on appeal; like Mr Davis, a millionaire businessman in Houston, Texas, charged with several murders and an attempted murder of his wife, like all those of wealth know: money talks.

Does anyone in America seriously think that Black juries don't convict Black defendants?

There are many doing long sentences, if not death sentences, across

Pennsylvania, Ohio, California and beyond who were sent there by predominantly Black juries.

There are thousands (if not tens of thousands) of Blacks in federal prisons across the United States who are there because mostly Black juries in the District of Columbia sent them there.

The published opinion of Simpson prosecutor Marcia Clark, to the point that 'Black juries won't convict Black defendants' is a damnable lie, damned by a record that shows otherwise.

The trial has proved something else: most Americans don't give a tinker's damn about 'reasonable doubt'.

The claim that 'reasonable doubt' is the common standard in American law is the legal equivalent of the lover's promise that he won't come.

It is a legal fiction that, this once, a jury took seriously; and witness the result!

If O J Simpson were a poor man of no influence, with the exact same errors in his case, he would still be on Death Row, for without money for good investigators, good jury specialists and good lawyers, who would know that such errors ever existed?

The O J Simpson case is so far from the average Black defendants' as to be, in effect, in another universe. Indeed, truth be told, the Simpson case is worlds away from the average white defendant's experience!

The angered, fevered reaction to the Simpson verdict reflects how deeply ingrained the notion of white supremacy is in American life.

It sticks in America's craw that a 'nigger' can be acquitted in the double-slash-killing of two white people — even a 'millionaire nigger'.

This goes against the grain of over two centuries of American custom and traditions, and 'wasn't supposed to happen'.

That it did happen rocked white America, and will be used as a cudgel to batter Black American interests for some time to come.

Ultimately, the defining element in this case will be seen not as black, nor white, but as green — the colour of money.

© *Mumia Abu-Jamal, 1995*

Mumia Abu-Jamal *is is an award-winning print and radio journalist. Despite convincing evidence of his innocence, he has spent the last 14 years on Pennsylvania's death row*

JUNE JORDAN

Big country, no nation

What happens on the East Coast makes 'national' news and determines the 'national' state of affairs. Viewed from California, says June Jordan, there is a different reality

IT MUST BE difficult to grasp what's going on in these United States. American mass media base themselves on the East Coast and, from there, blissfully categorise events, concerns and personalities as 'national', 'star', 'local', or unworthy of notice. But this is one big country! And if people in North Dakota spoke one language, while folks in South Dakota, or in New Mexico spoke another, a different, language, that would make sense, given the size of these various states, plus the climate, terrain and demography peculiar to each.

For example, California is so large you could travel the equivalent of the distance from Boston, Massachusetts all the way down to Richmond, Virginia, and, easily, still spend the night inside state borders. This is one big country.

The three hour time change from New York to California has meant that East Coast television pundits regularly 'project' the winner or loser of presidential elections long before a huge number of Americans living out here, on the West Coast, get to leave their 9-5 jobs and vote.

If a New York City college campus erupts with anti-Semitic propaganda, for example, then *The New York Times* may very well publish a 'national' analysis of such hatred and student polarisation, as the 'national' state of affairs.

This is one big country. And so I am forced to reject and dispute serious misinformation that postures as 'national news'. Apart from official statistics reporting the percentage of public monies allocated to education or to the construction of prison cells, 'nationwide' is too broad and too sloppy to mean anything.

After the jury verdict that set free O J Simpson, a 'great racial divide' became the national news image for relations between White and Black Americans. (Notably, the opinions of other kinds of Americans did not merit even perfunctory documentation.) However, the relevant statistics contradict the message of the metaphor: 30 per cent of White Americans believed that O J Simpson was innocent. In other words, 75-80 million White men and women supported the verdict of 'not guilty'. This means that twice as many Whites wanted Simpson back on the golf course than there are Black folks living here. The total African American population — including those who thought justice completely miscarried on the millionaire tracks of O J Simpson — the total African American population is 28 to 30 million. But, never mind these numbers! Post 'O J' and I found myself stuck in a 'Great Racial Divide'.

I decided to conduct a local reality check.

On the block where I live, there are Black families and White families, Southeast Asian Americans, Chinese Americans, 'interracial' students, elderly folks, newlyweds, Jews and Christians. This is a short little street of great good calm.

No racial divide in sight: perhaps I needed to get off the block, and broaden my perspective!

Nine days after the Simpson verdict I stood on my West Coast campus, half a mile from my house. That was 12 October 1995, when 5,000 Americans of every description jammed together in California sunlight for an affirmative action rally at the University of California at Berkeley. Posters and placards bristled the air with identity declarations. There were 'Queers for Affirmative Action,' 'Jews for Affirmative Action' and Native American dances for affirmative action. Everybody was invited and everybody showed up. As a matter of fact, the leadership for this fantastic success came from a new student organisation called Diversity in Action. As the name suggests, this gifted task force includes African American young men with dreadlocks or shaved heads and White

American young men with blond ponytails and young Chicanas and Vietnamese American young men and women, and like that, on and on.

This ecstatic enormous throng gathered together to demand restoration of affirmative action throughout the University of California system, and to assert an intelligent resistance to demagogic, racialised, un-American manipulations that would deny American history, deplore American diversity and ignore our obvious, principled unity.

Malcolm X, assassinated 1968: role model for Black militants

When I came to speak I pointed to the very recent (12 September) finding by the 1995 National Research Council report on the quality of PhD education in the US. With the most heterogeneous student population in the world, UC Berkeley is the leading, the top university in America or, as the *New York Times* reported, 'No other University even comes close.'

Our main speaker for this huge happening was the Reverend Jesse Jackson, and we greeted him with an endless tumult of cheers and excitement. He responded with a rousing argument in favour of 'real world' politics and policies and, therefore, affirmative action. He implored us to 'turn to' each other and 'not turn on' each other. He inveighed against odious visions out here: racism, sexism, anti-Semitism, homophobia. And he implored us to exploit the energy of our anger and vote Gingrich and company out of power in 1996.

That evening, national news briefly noted 'some campus demonstrations and a few arrests in California, earlier today,' before returning to further 'national analysis' of 'The Great Racial Divide.'

Four days later 'The Million Man March' (which produced neither a million men nor a march) took place (on the East Coast) in Washington, DC. A whole lot of Black men came to be counted, to count and to see what non-violent standing and talking, face to face, might open up.

I was not invited. I sat in my living room and I watched 'The Million Man March' unfold. But I have to admit some elements of prejudice that, undoubtedly, affected my perceptions.

• I have a bad attitude towards anybody or anything that wants to sunder me from the Black man who is my son.

• I have a bad attitude about the fact that Louis Farrakhan deemed Malcolm X 'worthy of death,' on 4 December 1964 in the Black Muslim newspaper, *Muhammed Speaks*.

• I do not agree that deliberately separating Black men from Black women is either a brilliant or an appropriate — a helpful — idea.

• I am always leery, at the very least, of leaders who tell the poor and the despised 'to take responsibility for yourself.' I become especially negative when, meanwhile, those leaders lobby for government monies for themselves. For example, since 1991, Nation of Islam affiliates have sought and secured more than US$15 million in federal contracts for 'security guard' services in public housing projects.

• It seems to me that 'atonement' by Black men is an okay ambition only assuming that everyone else on the landscape, White men, for instance, will cop to their crimes against Black men and Black women and everybody else suffering the consequences of White supremacist ideology.

• Between the vile hate mongering and the amazing megalomania of Farrakhan (who asserts, among other things, that 'God' speaks directly to him and through him) I am at a loss: which is more self-serving? Which is more destructive?

Nevertheless, I watched The Million Man March. (As a matter of fact, CNN-TV gave Farrakhan two and a half hours of *uninterrupted*

USA: STATE OF THE NATION

international outreach — way more than the Pope, Bill Clinton or President Nelson Mandela have ever received!) And I felt proud and I felt angry and I felt sad and I felt hopeful because it became obvious to me that, regardless of the actual or constructed leadership of that movement, my people, Black people again and again, move forward from a moral humility of spirit, a capacity for personal risk and humane good will that may yet prove to be our saving grace and strength.

As for the meaning of The Million Man March? Or, apart from 'national' interpretations and evaluations, what will The March lead to?

I have no idea.

But, on the afternoon following the March — on Tuesday, 17 October 1995 — our University of California faculty senate voted 124 to 2 to rescind the UC Regents' ruling against affirmative action.

While this vote does not assume that the Regents will reverse themselves, it does stand the faculty with the students who stand with the Chancellors, united against the Regents and for affirmative action.

The 'nationwide' assault on affirmative action began right here, in Northern California. And it looks to me like we may bury that particular outrage, right here, where it was born. For sure, the fight is on, and it does seem far from hopeless.

This is one big country.

I happen to live on the Pacific Rim, which, for better or worse, harbours the demographic and economic forecast for all of the USA, in the twenty-first century.

And I trust what I can see and what I can hear and what I can do on my block, and around the corner, and on the campus where I teach, half a mile away.

And, just now, I am awfully glad to live nowhere else but here: right here. ❑

© *June Jordan, December 1995*

June Jordan *is a poet and Professor of African American Studies at UC Berkeley. She is a regular columnist for the* Progressive, *and her most recent book titles include* I was Looking At the Ceiling and Then I Saw the Sky, *the libretto for a new opera written in collaboration with the composer John Adams, directed by Peter Sellars, and* Technical Difficulties

NETSCAPE

BRIAN WINSTON AND PAUL WALTON

Virtually free

From the printing press on, all new technologies have had their democratic potential firmly suppressed. The Internet proves to be no exception

ONCE upon a time there was a wide area network called the Internet. A network unscathed by the capitalist Fortune 500 companies and the like. Then somebody decided to deregulate the Internet and hand it over to the 'big boys' in the telecommunications industry. ... The Internet Liberation Front is a small, underground organisation of computer security experts. We are capable of penetrating virtually any network linked to the Internet — *any* network. ... Just a friendly warning Corporate America... ('Greetings from the Internet Liberation Front' reprinted in the *Guardian*, London, 12 July 1995).

Not since Eden, some would have you believe, has humankind been in such a state of grace as is now enjoyed by users of the Internet. In the new dimension of cyberspace, real freedom is at last possible. A virtual world of true democracy where no speaker is more powerful than any other has been created, unplanned and unsanctioned by the potentates of telecommunications and computing. A world, moreover, that appears to be, to all intents and purposes, free and unregulated. As the authoritative *Economist* recently put it: 'The growth of the Net is not a fluke or a fad, but the consequence of unleashing the power of individual creativity. If it were an economy, it would be the triumph of the free market over central planning. In music, jazz over Bach. Democracy over dictatorship.'

This is hyperbole. It is specious nonsense of a particularly disabling kind, in that it obscures and indeed prevents reasoned discussion about technological change and social policy.

The Net is a sort of computer-based hybrid between a telephone exchange and a broadcasting system. At one level it is nothing more revolutionary than a worldwide network of computers exchanging data telephonically, which they have done throughout the 50 years of their existence. Originally data was transmitted between the mainframes at the heart of the military-industrial complex and was about thermonuclear ignition problems. Now it predominantly takes the form of electronic mail transmitted between personal computers.

Given the level of hyperbole the Net is generating, let us remember Raymond Williams: 'In the early years of any new technology, it is especially important to clear the mind of the habitual technological determinism that almost inevitably comes with it.'

Technological determinism is that system of thought which simply seeks to assess technology in a vacuum ignoring all social, historical, economic and cultural factors. It is in fact a species of flat-earthism in that it concentrates on one set of phenomena to the exclusion of all others. The technohype over the Internet is a particular egregious example of technological determinism run totally amok.

In a world where, according to the International Telecommunication Union, half of humanity are more than a two-hour *walk* away from a telephone we need to remember, for starters, that we are talking about the Haves and completely ignoring the Have-nots.

Even the Information-Haves are not that blessed with this technology either. The commonly cited figure of 20 million Internet users seems to have little basis in fact. John S Quaterman, an Internet demographer from Austin, Texas, last summer estimated users worldwide at about two to three million. A recent survey suggested that Internet-connected computers in the UK numbered no more than 300,000.

And who are these users? According to the Georgia Institute of Technology, in the most comprehensive survey of Internet use to date (1994), 90 per cent are men, 80 per cent are white, 70 per cent are North Americans, 50 per cent spend 40 hours or more a week computing and 30 per cent are graduates.

This, of course, is no basis for denying the main thrust of the hype which predicts revolutionary impact for the Net in the future. However, the first task of any argument claiming profound effects for a new technology is to explain how and why it differs from previous technologies — not in its *technical* capacities and potentials (which is the failing of technological determinism) but in the social relations and cultural forms that have produced it. This the current hype makes no attempt to do.

In fact, there is, exactly, nothing new about the broader social circumstances behind this technology. The basis of the Internet is a system of telephone links (which dates back a century) augmented by the newer technologies of satellite coupled with digitalisation (which dates back to the '30s) and computing (which dates back to the '50s). The development of the Net is following established institutional patterns.

For example, in Internet mythology, the Net was supposedly created,

KIRSTY GORDON
Companions of the superhighway

Sex, exiled so long from the realm of technology, surfaces in cyberspace with all the symbolic resonance of the long-repressed. The Internet, presented as the democratic vehicle for the free exchange of (primarily academic) information, epitomises the technological achievement of civilised man: the proliferation of pornography in this medium, then, represents a failure of society's sublimating mechanisms to keep the primitive at bay. Is it surprising that its presence should provoke such a hyperbolic response? Cyber-porn has all the subversive impact of an unruly id breaking through the superstructures of the ego, and is as difficult to contain. The paedophile fantasies of one ill-socialised individual, code-name 'Blackwind', in Simon Winchester's 'An electronic sink of depravity', (*Spectator,* 4 February 1995), come to represent 'an almost exclusively American contagion', symptomatic of 'the barbarisms of the modern mind': without fixed origin, the discourse takes on the character of a collective utterance, emanating from an electronic analogue of the subconscious mind.

The threat posed by cyber-porn seems to lie as much in the medium as in the material itself. It is important to stress that in the case of alt.sex stories, no indecent act has been committed, no obscene photographs are proffered, and in an unplanned, unstructured fashion, by individual computer enthusiasts linking up their modems. But, on the contrary, it is a by-product of the growth of the world economy, a handmaiden of the transnational corporation. Far from being unplanned its main trunk line (the backbone) has been operated by the National Science Foundation of the United States government.

Yet again, this does not mean, of itself, the hype is wrong. Even if we allow that the Net's history is no different from other communications technologies whose democratic potential has not been realised, this still does not mean that the Net might not be different in the long-term. It is still possible that at last a truly democratic technology is at hand.

But when closely examined the case for what might be termed 'digital exceptionalism' rests primarily on the fact that the Internet was supposedly an 'unauthorised' and 'democra-

the pornographers are not interested in exploiting the money-making potential of cyber-sex: the pedaristic activities depicted have no corollary in the 'real' world. Pornographic fantasies of this sort have always existed: the Internet, however, makes them readily available to anyone with a modem and a mouse, regardless of age or cultural background. Those in favour of regulation argue the need to protect. Yet instead of pressing for realistic constraints — a more selective system of access to the dingier corners of cyberspace, for example (an electronic top-shelf, as it were) — the majority of the anti-cyber-sleaze crusaders advocate an impracticable censorship. It is the presence of pornography on the Net that is threatening regardless of who does or does not read it.

With alt.sex posited as the manifestation of depraved humanity, is censorship still an option? Is it even desirable to do so? The very existence of cyber-porn testifies to the futility of trying to keep technology distinct from human nature. The one reflects the other: in the same way that the dream of technological progress originates in the urge to transcend the 'barbarous' psyche, so extremes of sex and violence can be explored on the Internet precisely because, in this sublimated medium, sex and violence exist only as concepts — the physical is an impossibility. No longer a linear superhighway, the Net is developing into a complex entity that simultaneously abstracts and reflects the structure of human consciousness: surely, in this electronic hall of mirrors, there is a place for fantasy. ❏

Kirsty Gordon *is a freelance writer*

tic' application of computing and telephonic technologies. This contention (although clearly not based on the historical record) is in some way 'proved' by the fact that the Internet presents an apparently surprising absence of pricing mechanisms and enforceable controls. Cyberspace appears to be free.

But this is more apparent than real and is changing rapidly. In mid 1995, the National Science Foundation has handed the Net's backbone over to the private telecommunications giants Sprint, Ameritech and Pacific Bell. These will now become the gatekeepers, or principle access points. It was this development that caused the cyberadicals of the Network Liberation Front to announce themselves to the world. But even before this development, such Internet fundamentalists were in the grip of the strange delusion that they and their communications system stood outside of capitalism. Nevertheless, they had all bought computers, modems, software, subscriptions to online services and telephones.

Their illusion of getting something for nothing was entirely based on the fact that data transmission times have dropped so that it takes only two-thirds of a second, for example, to send an e-mail message from the US to Antarctica. Moreover, the Net breaks up even such super-fast messages transmitting them with scant regard to the time/distance cost structures of traditional telephone use. But, however fast and however efficient the routing, this is still not 'free'. The telephonic infrastructure is being paid for by users, but minimally. These costs become largely invisible because the Net itself is a very efficient user (and, indeed, abuser) of the infrastructure. To believe that the Internet is, in fact, free is exactly the same as believing that commercial television is 'free to air'.

The hype suggests, though, that the Internet is nevertheless too complex now to be controlled and priced because of the intrinsic nature of the technology. But how true can this be in the long-term? It is surely illogical on the one hand to claim this while at the same time denying that the same computing power cannot keep track of what is happening for pricing or, indeed, other purposes.

Handing the backbone over to commercial firms is only one of a number of reasons for supposing that the current invisible level of price will not long continue. Infrastructure pricing systems will be adjusted to take cognizance of the fact that data streams can be as valuable to individuals as they once were only to bulk commercial users.

So where does all this leave the Net's potential for democratic speech? Andrew Garton from Brisbane, Australia, argues that for a time it appeared the Net posed a 'threat to governments and corporations wishing to control information to their communities', but he was writing in autumn 1994 prior to its privatisation; and even prior to this he had noted the growing hostility of official bodies and their attempts to curtail aspects of network activity. He suggests that by the late 1980s the US had reached the stage, 'where the surveillance of certain individuals and confiscation of computer equipment, along with the closure of many electronic bulletin-boards (micro-networks often used for specialist information sharing), resembled the red scares of the 1950s'.

He cites the US operation 'Sun Devil' of May 1990 which involved 28 raids in two weeks and the confiscation of 42 computers and 23,000 discs. And presently there are moves by the Clinton administration to regulate the use of encryption software to ensure the security of daily exchange. Researchers have argued that as the backbone is located in America, electronic mail is 'wide open to US surveillance'. Also, several people last year have been charged by the British police for abuses on the Internet mostly relating to pornography, fraud or pederasty. Yet a study reported by T Mathews reveals that significantly less than one per cent of computer traffic on the Internet relates to pornography and violence.

Pressure from the state on Internet service providers has led to Internet connections to users being cut in both the US and Canada. By mid-June 1995 America Online — the largest single commercial provider for message senders on the mailing lists which were examined — was reportedly cutting off half a dozen users a day for 'Net abuse'.

This is not to say that the Net is entirely conservative in its communication. There are networks such as the global Association for Progressive Communications (the APC), which have been providing low-cost Net access to over 100 communities both in the developed and developing world. Andrew Garton reports that during the Soviet coup of 1991 the Russian staff of the Moscow-based GlasNet opposed to the coup provided the most direct and immediate reports of what was happening. Susan O'Donnell has recently studied activist groups on the Internet and was able to show that 14 per cent of messages were linked with solidarity actions outside the Internet and around 10 per cent of all messages circulating on such lists encouraged readers to participate actively in joining such groups.

Despite such findings, since the early 1990s a whole range of US groups including the Electronic Frontier Foundation, The American Civil Liberties Union (ACLU), and other First Amendment supporters have had to struggle against the tendencies pushing control and privatisation. It is clear that the state authorities can monitor exchanges. And not just the state. For instance, newspapers report that in the UK private security forces had joined GreenNet for the express purpose of monitoring the behaviour of activists working on anti-roads campaigns.

The point is that all communication technologies from the printing press on have democratic potential. History reveals that these potentials are normally suppressed. The danger with the current hype over the Internet is that because it looks simply at technological possibilities it directs attention away from what is actually happening. Freedom on the Internet is not inevitable. It is rather, as ever, a question of political struggle.

It is no accident that the ideologues of the *Economist* and leading politicians such as American Vice-President Al Gore want the hidden hand of the free market to dominate pricing and access. This, of course, suits the transnational corporations but the result is not then a democratic Information Highway. It is, rather, an Information Toll-Road — and it looks increasingly likely that the toll-gate keepers will not only charge for travel but also insist on examining every last piece of baggage. The cyberadicals of the Internet Liberation Front are going to have their work cut out for them. History is not on their side. ❑

Brian Winston is director of and *Paul Walton* is senior research fellow at the Centre for Journalism Studies, Cardiff University

Once and future shock

Ten years on, the real tragedy of Chernobyl is that the disaster is still growing, often in secrecy, and is far worse than anyone predicted. The humanitarian, health and economic problems are overwhelming the states affected — and there are more Chernobyls waiting to happen

(Left) Inside Chernobyl 1995: still counting the cost
Photo by V Kormilkin

ANTHONY TUCKER

Annual Chernobyl march, Minsk 1992: 'Belorussia. As one with Chernobyl'

Confusion and deceit

For those unaffected, memory of disaster is short. Who, in the well-heeled West, now recalls that the world first learned of the Chernobyl accident from Swedish scientists and American spy satellites? Or that Gorbachev, promoter of *glasnost* was, like the world beyond Soviet borders, kept in the dark about the enormity of the accident and the irradiation problems his country faced; or that the period of panic when fallout rained down on western Europe was accompanied by confusion and disinformation, that scientific reports from affected areas were fragmentary, filtered, impossible to validate? Or that, when the brave 'sarcophagus', holding a lid on the grave of the shattered reactor became in serious danger of collapse and Professor Velikov, former chief adviser to Gorbachev, came to a Pugwash meeting at the Royal Society in London to beg for help, there was no response?

Sure, the early silence, confusion and Europe-wide panic led to massive pressure for the speedy international notification of nuclear accidents involving trans-boundary effects, although it is not clear what such an agreement might achieve. Preparedness in the West has barely improved: changes mirror the politically correct message that Chernobyl cannot possibly happen here.

In the Soviet Union the early desperate and often incredibly courageous measures to control the monstrous meltdown, limit damage, and evacuate large areas, were carried out as a huge military operation amid a morass of technical and political confusion. Tragically for affected civil populations, this confusion was transformed into chaos as the Soviet Union itself burst apart, fragmenting into the new (or rather old) states whose pride, sensitivities and separate health ministries shattered all immediate hope of co-ordinated follow-up and remedial programmes.

Russia, a nuclear state whose other Chernobyl-type reactors are still essential to power supplies, had a powerful political need to bury the Chernobyl image as quickly as possible. Worldwide cynicism and distrust were inevitable. But the real tragedy of Chernobyl is not that its immediate consequences and immense long-term problems are fading from memory, minimised, played down, obfuscated by the governments and protection agencies of the former Soviet Union and the western world, but that as a disaster it is still growing and that in many ways it is far worse than anyone predicted.

A stark and emotional report by the secretary-general to the fiftieth General Assembly of the United Nations[1], has revealed that, a decade after the immediate fallout, the humanitarian, health and economic problems entrained by Chernobyl are still emerging and still growing. They are already so huge that they are overwhelming the affected states and threaten to overwhelm the comprehension, the compassion, even the imagination of the world.

The lives and livelihoods of around 10 million people have already been affected. Half a million people have been displaced. Predictably, the

ANTHONY TUCKER

abandoned villages and forests of the 30 kilometre exclusion zone around Chernobyl have become a wild, sinister, no-go haunt of criminal and bandit communities. But in Belarus, in the Russian Federation and in the Ukraine, where weather determined fallout would be greatest, agriculture is corrupted by contamination, there is massive social and industrial dislocation, and humanitarian, health and economic problems are of such immensity and complexity that they are far beyond available resources and are perhaps comparable only with the aftermath of civil war.

Ideals are largely pipe-dreams, but even in a world in which we expect only that international responsibilities will be taken seriously, common sense dictates that all nuclear governments should have worked closely in the provision of support, equipment and expertise to investigate and remedy Chernobyl consequences. This is not a matter of generosity or of international aid, but of practicality. Chernobyl offered and still offers the world an opportunity to gain a unique understanding of the effects of nuclear disasters, as essential for critical awareness as to provide a real basis for future emergency responses.

Of course, few countries can afford, like the former Soviet Union, to entomb as radioactive debris a thousand helicopters and the several thousand buses needed to evacuate large populations from contaminated areas. Yet all countries need to know, through independent and openly published investigation and skilled technical interpretation, the true detail of what has happened and what will happen in the decades ahead to the millions of relatively poor and uninformed people whom circumstances force to live amid contamination that will remain for generations.

It hardly matters now that, as it wove its way round Europe and eventually round the world, the Chernobyl plume produced an immediate and powerful amplification of existing public hostility to nuclear power. Official statements were dismissed as deception: the Swedish public demanded an end to nuclear power, the French were evasive, the Americans kept their heads down and the British government quietly delayed the go-ahead for Sizewell B for a few months. Then it was back to business as usual.

THE GROUND for this had already been laid. In the immediate aftermath amid wild predictions of huge casualties and following correct protocol, the International Atomic Energy Agency (IAEA) — to which nuclear governments belong — produced a large and detailed report. It was based primarily on Russian information and was unexpectedly reassuring, suggesting that overall consequences and long-term effects within the Soviet Union would be much less serious than many feared and, outside the Soviet Union, any effects would be negligible. At the same meeting in Vienna, at which no independent data were presented, the IAEA sought an immediate safety assessment of all Russian reactor designs and proposed guidelines for the assessment of dose, continued monitoring and follow-up of the reactor workers and the emergency teams most directly affected by radiation.

Although it can conjure up design expertise from member states, the IAEA has little independent money for the support of huge programmes of technical investigation and, in matters of health and epidemiology, plays a specialised but strictly limited industry-focused role. Protocol requires, properly, that the World Health Organisation at Geneva (WHO) should be the executive agency for international health programmes, although WHO is itself continually starved of funds. This means that promotion of the vast, specialised and long-term medical programmes of the kind demanded by Chernobyl is largely dependent on dedicated additional finance from member governments, agencies or bodies like the EC. Even then, neither WHO nor the IAEA can act as an independent investigator in the collection and publication of potentially sensitive medical, scientific and technical data.

> All countries need to know the true detail of what has happened and will happen to the millions of poor, uninformed people whom circumstances force to live amid contamination that will remain for generations

Neither has the power to enforce collaboration and participating governments filter their own information. Additionally, in the Chernobyl follow-up, the affected governments chose to channel data only through WHO and the IAEA, thus excluding all normal peer-reviewed routes for scientific and technical publication.

Only SCOPE — the International Union's Scientific Committee on Problems of the Environment — was able to set up an independent programme, and this was limited to monitoring and assessing models of the physical and biological pathways of radioactive elements in different environments.

Thus, although on the face of things it may have seemed that the international community had joined hands to provide the resources, the expertise, the training and the planning needed to cope with and to learn the important lessons the disaster could deliver, the programme was hedged round by crucial limitations. Although expert groups from western Europe and the USA visited their academic and medical colleagues in the seriously affected regions, they did this on their own. The only hard money was a dedicated sum of US$20 million from Japan.

This set WHO's Geneva-based IPHECA (International Programme on Health Effects of the Chernobyl Accident) in motion. After three years, neither acceptable study protocols nor useful data had emerged. The money had vanished in new equipment, mainly in Russian laboratories. Western experts, acutely aware of this failure of management and of the poor lines of communication, advised that IPHECA was itself a disaster. Pressure for reform came, not from WHO HQ at Geneva, but from EC groups collaborating in Belarus and from WHO(Europe) in Rome, whose radiation arm had developed studies alongside the Belarus Ministry of Health.

At this time, little science of western standard was emerging from the affected areas. Worse, the highly prestigious scientific journal *Nature* had been accused of censoring Chernobyl fallout information by accepting papers from western groups and then failing to publish them, and Britain's National Radiological Protection Board stood accused of failing to correct misleading undermeasurement of fallout over Britain, published in its first scientific report. It remains uncorrected. Then, as EC and WHO(Europe) collaborative studies began to pick up evidence of a massive fallout-related increase in the incidence of thyroid cancer among the children of Belarus, their findings came under attack from 'official' Russian-dominated IPHECA scientists (who sought causes other than fallout) and from the United States.

While, in the early stages, the absence of organised studies could be put down to confusion and excused as part of the general cock-up, it was suddenly painfully obvious that western experts collaborating in Belarus and elsewhere could expect obstruction coupled with political and professional attack. Transfer of the essential science and skills in pathology and epidemiology, crucial to the future of large and suffering populations, was being blocked. This was not cock-up: it was cover-up.

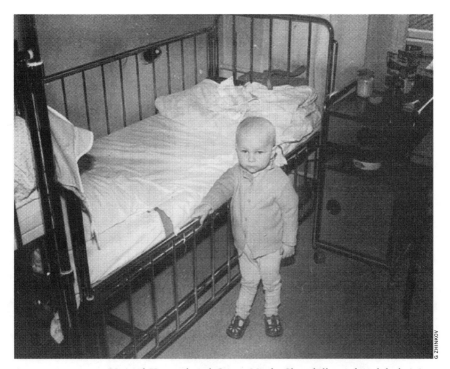

National Haematological Centre, Minsk: Chernobyl's unacknowledged victims

Nuclear states, trapped in the same schizophrenic dilemma whatever their colour, sought future safety in all honesty. Yet they also hoped that Chernobyl could be buried in blurred statistics and contradictory arguments, from which its legacy could never be convincingly unravelled. Even though known doses from fallout tellurium and radio-iodine were large and certain to increase the incidence of child thyroid cancer, politics required that studies confirming this increase must be attacked.

The US government had, and still has, powerful political reasons for obfuscating these findings. Its Department of Defence is inundated by civil compensation actions from US citizens seeking damages for thyroid cancer and other health detriments, claimed to result from exposure to radio iodine and other fallout around the Hanford reactors and from weapons test sites in Nevada and elsewhere. (Over the years Hanford has released almost one million Curies of radio iodine — about 12 per cent of the Chernobyl release.) Beyond its borders, thyroid cancer excesses had also started to emerge among Pacific Islanders affected by fallout from US nuclear tests. On top of this, some parts of the US nuclear medicine profession feared that their profitable freedom to use radio isotopes might become subject to new restrictions.

Whatever the appalling health consequences emerging among the exposed populations of Belarus, the Ukraine and Russia, the confusion and inadequacy of programmes, as measured by western standards, still acts as a double deterrent in the countries most able to help. In many the internal political position on Chernobyl is best served when scientific and medical reports reaching the West are of such low technical quality that they can be dismissed.

One disgrace is that the world's discredited radiation protection community remained and remains silent. Fortunately, on a narrow front, truth is winning. A Belarus study of thyroid cancer has been taken over by WHO(Europe), new programmes are planned (but not financed) and the latest data are irrefutable. These imply that among the infant group (nought to three years) exposed in the highest fallout areas, up to 25 per cent can be expected to develop thyroid cancer. One in four or five, instead of one in a million.

Even in isolation these figures are horrendous, implying the need for increases in medical resources and surveillance that lie well beyond anything available in the affected states. Worse, they are only a tiny fragment of the huge, interwoven and still emerging humanitarian problem fermenting in the wake of Chernobyl. The importance of the report from the secretary-general of the United Nations is that it cannot be ignored. It signals formal international recognition that the problems entrained by a nuclear accident are vastly greater and cover a far wider range than the health detriment measured in future cancers, still officially seen as the major criterion in assessing the effects of radiation.

It raises some uncontroversial but salutary questions. For example, why has there been no systematic follow-up of the army of Soviet 'liquidators', almost 200,000 strong, each suffering 90 seconds of intense irradiation inside the hulk of the Chernobyl reactor as they saved the world from far worse consequences? They are now dispersed, many untraceable, yet they have become a dramatic source of anecdotal evidence implying very high levels of morbidity and mortality (page 94). But what of the future of the huge agricultural populations forced, by economic necessity, to continue to live and work on contaminated land, or those who, as families, chose displacement and desperate poverty as the lesser risk to their children? With agricultural areas in Belarus and the Ukraine the size of England and Wales so heavily contaminated that they cannot be worked until well into the twenty-first century, there is crippling dislocation. International economic and humanitarian aid is needed as urgently by these countries as by any nation stricken by famine or by natural catastrophe.

The Russian Chernobyl Committee figures, seemingly endorsed by UNICEF in the report, suggest that there are around eight million seriously affected victims of the Chernobyl aftermath in these regions alone. This may well be right. It may also be right to point to the observed and often large increases in a wide range of diseases in these areas as a product of stress, an issue judged to be important and real by an expert group at WHO(Europe)* in 1994.

Sadly, the UN report is open to dismissive attack. It quotes such low pre-Chernobyl disease incidence figures that they reveal past under-ascertainment, implying that apparent increases are merely a product of better surveillance. Its reports of stress-related increases in morbidity, birth defects and leukaemia are largely anecdotal and, although possibly real, have no firm statistical base — a weakness already exploited in knock-down comment in the nuclear establishment literature[2]. Although powerful, the UN document will face the same scientific hostility as the early reports of epidemic child thyroid cancer in Belarus.

Yet this turned out to be real and, since there is increasing evidence that psychological stress adversely affects the immune system[3], the increases in morbidity may also turn out to be just as real and disastrous as the UN report suggests. In any case, the economic and human disaster is clearly of such proportions that, like famine or catastrophic earthquake, it evidently merits international aid on a massive scale, free of nuclear hostilities and, above all, properly co-ordinated. Is this another pipe-dream? ❏

Anthony Tucker is a former science editor of the Guardian, UK

[1] UN Document A/50/418 of 8 September 1995
[2] *Journal of Nuclear Medicine* 1995; 36,9, p29N
[3] eg. *Lancet* 1995; 346: pp1194-96
* The author was a member of this group. Its draft report has been circulated but awaits agreement.

MIKHAIL BYCKAU
A liquidator's story

For the first time in print, a Belarusian scientist gives his personal recollections of the secrecy that, in the crucial period immediately following the Chernobyl accident, left the unsuspecting public exposed to fallout

ON THE MONDAY morning, 28 April, at the Nuclear Energy Institute of the Belarusian Academy of Sciences, I switched on the apparatus — the gamma-spectrometer and the dosimeters: everything was (in physicists' slang) 'hot', which meant that there had been a big nuclear accident on the Institute's premises: our dosimetrist ran out of the laboratory, and reported that the level in the yard was about 300 microroentgens an hour. Then he was summoned by telephone to monitor the radiation contamination round the nuclear reactor of the Institute of Radioactive Technology; so that was the main source of the accident! But they had their own dosimetrists there, and the dose level was almost the same; the same was true in the vicinity of a third nuclear device... Moreover, it was clear that the radiation levels fell the further one went inside the building... When the head of the dosimetry service, A Lineva, telephoned the Central Public Health Station of Minsk, they said, 'This is not your accident.'

We looked at the tall smoke-stack, and then at the map of Europe, and we saw that the wind was blowing radiation towards Sweden. In fact, we learned later, on 1 May the level of radioactive contamination in Stockholm was 17 Curies per square kilometre from Caesium-137, and 87 Curies per square kilometre from Iodine-131).

But in our place, they brought me in a twig from the yard, and I observed that it was emitting radiation...the gamma-spectrometer showed Iodine-131 and other 'young' radionuclides... Later we tested soil and trees from many regions of Belarus, and the Institute started to measure the specific activity of foodstuffs arriving for the Institute canteen and the crêche.

Meanwhile, the dosimetry service headed by M V Bulyha was monitoring the radiation cloud hanging above Minsk.

We started to ring our relatives and friends in Minsk, advising them about safety measures. But this did not last long: at around midday, our telephones were cut off. And a couple of days later, we specialists were called into the

Secrecy Department, and made to sign a 29-point document forbidding us to divulge secrets connected with the accident at the Chernobyl-plan. These included the structure of the RDMK-1000 reactor, the amount of uranium, etc, 'secrets' that had already been published in scientific literature.

And meanwhile out in the street, radioactive rain was falling...

We went home from work without looking from side to side; it was painful to see how the children were playing in the radioactive sand, and eating ices.

In our street, I went up to a street vendor and told her to stop selling her sausages, as radioactive rain was falling. But she just said:'Be off, you drunkard! If there'd been an accident, they'd have announced it on radio and TV.' A naive soul, she believed in the righteousness of the Soviet authorities.

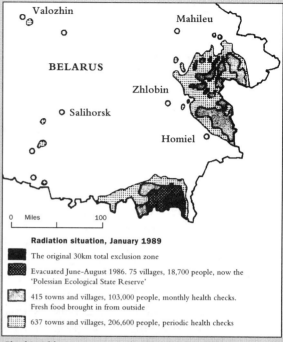

Shuskievich's surprise: map released 1 February 1989 (see page 114)

In the evening, on Central TV, Moscow showed us how tractors with great swirls of dust behind them were tilling the soil down in Naroula country, part of which lies in the 30-kilometre zone around the Chernobyl station. Then, on 1 May, as always, children and adults marched in columns through the streets without even guessing at the consequences. So now, today, in Belarus we have some 400 children with thyroid cancer...who at that time knew nothing about Iodine-131...

Mikhail Byckau *is a nuclear physicist, who from mid-May 1986 until his retirement from the International Sakharov Institute of Radioecology in April 1995, played an active role in the 'liquidation' (clean-up) and monitoring programmes in the contaminated areas Translated by Vera Rich*

DAVID HEARST

Taming the beast

Co-operation, not competition, is the only way the West can contain the threat of Russia's volatile nuclear industry

TEN YEARS after Chernobyl, Russia's nuclear industry is wary of western admonitions about nuclear safety. They are sceptical, too, of attempts to stop them selling light-water reactor technology to Iran. And they have every sympathy with Armenia for restarting its reactor, close to an earthquake fault-line. The nuclear industry, once fully closed to world view, is more open than it has ever been to outside inspection, but in many western offers of 'help', Russia now senses a different agenda.

Citing 'military sensitivity' Russia has so far barred US inspectors from a full safety inspection of Krasnoyarsk, one of two remaining reactors producing weapons-grade plutonium. The USA has spent US$27 million to counter dangers from nuclear waste in the former Soviet Union, largely from production of weapons-grade plutonium. The General Accounting Office, which prepares reports for the US Congress, cited a private report suggesting that Russia's nuclear weapons programme releases 650 times as much radioactivity as the US one.

The US is also involved in programmes which do not involve the inspection of military-sensitive sites. For example, US computer experts are teaching Russia's nuclear scientists how to introduce a secure system of computer tagging. This involves transferring a paper inventory of millions of pellets containing enriched uranium on to computer, ending the paper nightmare in which no plant director can say exactly how much uranium he has, and where it is. The argument as Congress sees it is simple: why does Russia appear to bite the hand that feeds it?

The answer is less complicated than the West would imagine. Nationalism feeds on economic and political decline, but Russia's nuclear industry is more sophisticated than that. Alongside the appalling safety record and history of accidents and cover-ups, it developed technologies, like centrifuge developed to enrich uranium, far in advance of western chemical methods.

It also developed micro-reactor technology for launch into space. The industry's riches were not in the ageing, corroding plant, but in its brain power. These brains were the elite of the powerful military-industrial complex, the children of Stalin's feared NKVD chief Lavrenti Beria. Today these scientists have to grow their own food to supplement their income which, like all other state salaries, arrives late. In real terms, their salaries have fallen to levels that are ludicrous in relation to their awesome responsibilities.

The director of a fast-breeder research reactor, told me, on condition of anonymity: 'Even five years ago with my salary, I could put petrol in my car, go to Moscow to see a play in a theatre, go out to a restaurant and then come back. All that is finished now. I can't even permit myself to use the car as before. With my 300,000 rouble (US$70) salary, a tankful of petrol is a luxury for me.'

The Russian nuclear complex still has its secrets — but they come much cheaper to any potential buyer. The first four years of post-Communism has done the Russian nuclear industry no favours. It was one of the first victims of the western-sponsored shock therapy reform programme, when prices were liberalised, state subsidies to industry slashed, when the spiv street trader found he could earn more than a nuclear researcher.

Partial de-industrialisation was not a casual by-product of western economic advice, but one of its policy objectives; so was making the division of the Soviet Union irreversible. In the post-Cold War triumphalism of the West, market reforms seemed such a good way of slaying the Soviet military-industrial beast, replacing it with a civilian but weaker industrial profile, akin to a third world country.

Russia's nuclear industry, formerly disingenuously known as the Ministry of Medium Machine Building, was an integral part of the military-industrial complex. Demoralised by Chernobyl, whose scale still had not been officially admitted, the nuclear industry was an easy target for the radical democrats in Russia and nationalist regimes in the former Soviet Union. All new construction work was halted. Salaries were frozen. The industry had to fight to justify its existence.

In Ukraine, Lithuania and Armenia nationalist movements came to the fore on a platform of closing down former Soviet reactors, as symbols of Moscow's nuclear-political tyranny. In Armenia, an orchestra accompanied the closure of Metsamor, the light-water VVER-230 reactor that had survived the 1988 earthquake.

Yet now Nikolai Rabotnov, deputy director on Fundamental Research, of the Institute of Physics and Power Engineering, Obninsk, says: 'Each of these republics is doing everything it can to keep these nuclear facilities open. In Lithuania, there is Ignalinskaya Plant which produces 87 per cent of the republic's electric energy. How did they solve the problem? By inviting Minatom specialists from Russia, offering them salaries and housing just to maintain the station. Look at Kiev. They can't close the Chernobyl station, because every kilowatt per hour is precious for Ukraine.'

Six bitter, energy-starved, winters later, one reactor at Metsamor, near Yerevan, has re-opened. The president is still the same and so is his nationalism. But the lights are back on again — for a few hours at least — in a small republic crippled by its long war with energy-rich Azerbaijan, over the mountainous Nagorno-Karabakh enclave.

Throughout the former Soviet Union, the cry has been the same. 'Give us democracy please — but with central heating.' The rupture of economic and energy links between Russia and its former satellite states, soaring energy prices and the collapse of industrial output all made sick economies even more reliant on nuclear energy. No more so than in Ukraine, which is now involved in a wrangle with the G7 over money.

Not without reason, Ukrainian Prime Minister Yevhen Marchuk is saying, 'Put up or shut up.' The G7 has offered US$2.25 billion in grants and credit to close the station and supplement the lost generating capacity. Ukraine initially sought US$4 billion, but even that now is probably an underestimate, with the nation unable to pay soaring Russian gas and oil prices, on which the eastern half of Ukraine is entirely dependent.

Marchuk is left with only one argument with the West: 'Ukraine will be obliged to take a decision on modernising at least two reactors of the Chernobyl station, if the West does not offer serious money.'

The West's response to these arguments has been mixed. The USA first came to recognise the contradiction inherent in their post-Communist policies when it came to nuclear disarmament. It soon became clear that Kazakhstan and Ukraine could not deal with their post-Soviet inheritance of strategic nuclear warheads without direct and substantial US involvement.

'Prohibited area' 1993: one of the 'hot spots' that remain

In Russia the problem of decommissioning nuclear warheads is much greater. Under Start 1 and Start 2 (if it ever gets ratified) Russia is bound to dismantle 8,500 warheads from its strategic long-range missiles and bombs by the year 2003. Further commitments on intermediate and short-range missiles in Europe have placed another 2,000 warheads on the growing decommissioning list — that is, 10,000 warheads. The process is slow and costly.

Under a US Nunn-Lugar programme, 2,800 former Soviet warheads have been removed from Russia, Ukraine, Belarus and Kazakhstan. Congress has appropriated US$1.3 billion for the programme and US$75 million is being spent to build secure storage at Mayak and Chelyabinsk. But the programme is vulnerable to attack from Republicans who turn back to Cold War logic. Why, they ask, should we be helping an unstable Russia, who could turn around and elect a hostile, nationalist president in six months time? Under their pressure, Congress has restarted a project to develop, but not deploy, a multi-site Anti-Ballistic Missile defence system which Russia has warned could restart the arms race.

DAVID HEARST

The Russian political response is clear, and the pro-US president, Boris Yeltsin, is quite clear about this. Any new US AM system could mean the abandonment of Start 2, and the whole Russian programme of destroying its multi-warheaded ICBMs.

The West is now uncertain how to deal with Russia, its nuclear industry and its military might. It still has 27,000 nuclear warheads. It still has huge nuclear potential. The former policy of prising Moscow away from Yerevan, Kiev, Alma-Ata and Minsk, is, in nuclear terms, counter-productive. Minatom specialists are back in there, because these republics are even weaker than Russia, and need the electricity even more. Worse still, Russia has a huge problem, emerging from its militarised past — the quantity of weapons-grade plutonium that decommissioning is producing.

Some western governments have a more sophisticated response than others. Instead of trying to stop or curtail Russia's nuclear industry, Siemens of Germany, Bayernwerk AG of Bavaria and Russia's Atomic Energy Ministry have announced plans for construction of a new-generation nuclear reactor at Sosnovyy Bor near St Petersburg. The reactor is new generation of the VVER light-water reactor, the VVER-640, whose operating life will be 50 to 60 years. In policy terms, there really is no other choice but to co-operate in partnership, rather than in commercial or geo-political competition, with Russia's nuclear industry. The West can't slay the nuclear beast it once feared. Russia still has 27,000 nuclear warheads. The West can, however ensure that they do not fall into the wrong political hands. ❑

David Hearst is Moscow correspondent for the Guardian

ÜMIT ÖZTÜRK
A little radiation does you good

Tea house in Turkey

FOR A FULL WEEK after the disaster in Chernobyl, Turkish journalists, still smarting after five years of military rule and with warnings from the top to go slow on the story, confined themselves to reproducing what they read in the foreign media. General-turned-President Kenan Evren, along with recently-elected prime minister Turgut Ozal, urged 'calm' on the population and discouraged any investigative reporting by the local press corps on the grounds that Chernobyl was a long way from Turkey and the country ran no risk of contamination.

On 5 May, however, the foreign media reported the arrival the previous day of nuclear clouds and radioactive rainfall over the Black Sea region in northern Turkey. Red-alert headlines in the national press whipped the nation into total panic; with no professional advice or guidance from the top, the population developed its own prophylactic measures for dealing with radioactive contamination: well-roasted meat and milk filtered through women's headscarves, they argued in the villages, would minimise the risk.

After weeks of public outcry, Cahit Aral, minister for industry and trade, finally made a public statement in which he officially denied any danger of contamination. Ahmet Yuksel Ozemre, director of the Turkish Atomic Energy Institution (TAEK) and later to become a leading figure among Turkey's nuclear industrialists, followed suit, confirming that there was no risk and urging people to 'eat everything; eat whatever you find, without fear.' Meanwhile, Turkey's tea and hazelnut crops, the main source of income in the Black Sea region, were ready for export.

Critical comments in Turkish dailies by a group of scientists challenging the government's figures on contamination met with a firm response from Minister Aral on 24 June: 'Anyone claiming that radiation [from Chernobyl] has affected Turkey is an atheist and a traitor. Why sabotage our tourism and trade in this way?' The mainstream media was cowed into silence but the UK sent back its imported Turkish tea after discovering radiation counts well above the norm. In September, Germany and the Netherlands followed suit with the hazelnut crop.

Ensuing months saw a wicked farce played out with the collusion of the Turkish press. Political leaders from the president and prime minister down posed on its pages drinking tea alongside headlines such as 'A little radiation does you good.' 'I drink seven or eight cups of tea a day. Even 20 is harmless... We never dispatched any contaminated tea for sale... Even my wife drinks tea daily,' said Cahit Aral. 'Radioactive tea is more delicious, more tasty,' urged Turgut Ozal. The tabloids went even further, asking: 'Does a certain level of radiation have an aphrodisiac function?'

Disturbed by the supine compliance of most of their colleagues, certain journalists like Sukran Ketenci of the leading daily *Cumhuriyet* and scientists like Dr Inci Gokmen, assistant professor at the faculty of chemistry in the Middle East Technical University (ODTU) in Ankara, eventually got to grips with unearthing and publicising the facts concealed by government and press. The political monthly *Bilim ve Sanat* (Science and Art) devoted a special issue to the impact of Chernobyl in Turkey. It included pieces by the dissident professor of nuclear energy Tolga Yarman and the journalist and academic Haluk Gerger. Gerger called on the government to recognise the 'right of people to access to information on the effect of Chernobyl on Turkey'. Meanwhile, Dr Gokmen leaked the results of the counter-analysis by dissident scientists showing the radiation levels in tea, milk, hazelnuts and other commodities to be much higher than those given by official sources.

Aral responded by redoubling his reassurances in Parliament and in the press. He also asked Gokmen and her colleagues to go to TAEK, redo their analysis on the equipment there and discuss things with Ozemre. Ozemre himself wrote to Gokmen's rector demanding her dismissal along with her colleagues on the grounds that 'their equipment is totally useless, they are not qualified, they are distorting and manipulating the figures and their intentions are not good.'

At a prolonged meeting at TAEK on 13 February 1987, the scientists failed to persuade officials to allow them to disclose more of their results and were put under pressure to sign a ready-prepared statement endorsing the official line. In the face of continuing government denial and accusations of scaremongering in the press, Dr Gokmen and her colleagues persuaded the popular daily *Hurriyet* to publish their findings on 27 January 1987. Noting that the radiation level of tea was well above official figures, their report called on the government to initiate a campaign to raise national awareness of the risks involved; to warn children and pregnant and breastfeeding women not to drink tea at all; to stop marketing the radioactive tea; and to eliminate all remaining stocks.

All efforts to inform the public met with a concerted campaign orchestrated by the government to hush things up. In May 1987, the

Çernobil 1986   Akkuyu ????

Campaign poster against nuclear reactors in Turkey

government set up the Committee for Radiation Safety, under the chairmanship of Aral, ostensibly to investigate the impact of Chernobyl. As the minister responsible, Aral used the committee to spearhead opposition to the scientists and effectively stifle any public debate.

The media was only too relieved to move on to less contentious matters. The last word was with Minister Aral: 'Pre-washing eliminates the risk. Even my wife pre-washes the tea before she boils it.'

And there matters rested until the end of 1992, when now ex-Minister Aral blundered into a confession during an informal press briefing. The government had indeed 'hidden the facts and figures of Chernobyl's impact on Turkey', but, he added in a bizarre attempt at justification, 'We did take our revenge on the Soviet Union for the nightmare of Chernobyl by exporting the contaminated hazelnuts to Russia.' He also confirmed that the same nuts — a luxury item in the average Turkish diet — had been distributed free to Turkish soldiers. ❑

Ümit Öztürk is a Turkish journalist working on environmental issues, currently living in the UK

GRAPH 2nd Series

The controversial cultural journal **Graph** grew out of a profound unhappiness with the state of criticism and reviewing in Ireland. **Graph** provides a forum where lively engaged writing examines the intersection on culture and development, literature and science, politics and language. This second series reins in the talents of two new editors, **Bríona Nic Dhiarmada** and **Evelyn Conlon**, who join the original group of **Peter Sirr**, **Michael Cronin** and **Barra O Seaghadha**. **Issue 1** published in September included contributions from **Seán Dunne** on the Irish poetry business; **Anne Fogarty** on Irish women poets; **Michael Cronin** interviewing Murray Gell-Mann; **Seamus Heaney** and **Stanislaw Baranczak's** translations of Polish laments; **Fintan Valley** on 'communitas' of traditional music, **Aingeal de Búrca** ar na straitéisí léinn a bhí i réim maidir le traidisiún na Gaeilge ó aimsir na hathbheochana ar aghaidh; **Robert Fisk** on Kate O'Brien. **Issue 2** to be published in March includes an interview with **Ciaran Carson**; articles by **Louis de Paor** and **Michael Cronin**; a round table discussion with African Writers; **Ruth Riddick** on Women and Film Culture and **Lionel Pilkington's** Reconciliation Culture: Myths of Cooperation in Northern Ireland.

Price: £15.00 Spring and Autumn ISSN 07910-8016 210 x 146mm 139pp illustrated
Also available from all good bookshops at £6.50 per issue

BABEL

Children of Chernobyl

The following excerpts and illustrations from *Footprint of the Black Wind* (Minsk 1995), were produced by Belarusian school children in response to a competition on the theme 'Chernobyl in my destiny'. The writing by children from within or near the 30 kilometre exclusion zone expresses their confusion, grief and sense of loss at their evacuation from their homes. Children from further afield, in areas officially declared safe, discovered only three years later the true extent of the disaster and the damage to their lives

WHEN THE ACCIDENT happened, I was seven years old. I was in the reception class at Komarin school. I don't remember that day very well. The people of Komarin were very excited. Someone was running off somewhere, someone else was phoning. My sister and I kept hearing the words, over and over. 'An accident! Is it dangerous?' But Mummy and Daddy didn't pay much attention to this. Children went on playing in the streets, lots of people went out into the fields to plant potatoes. It was very warm. Then there was a little rain and we wanted to run around in the fields. Suddenly, they cancelled the May Day parade. Then we all got scared. And I often saw Mummy crying, and father and Grand-dad started talking about some kind of evacuation.

And all at once an unfamiliar word started to go round from mouth to mouth: 'Radiation'. A panic started. Everyone was trying to get away from Komarin, or even right out of Belarus, as fast as they could. It was just like the War. Time was suddenly divided into two unequal parts — before 26 April and afterwards. I remembered films about the War: people running, over-crowded railway carriages, a terminus swarming with people — everywhere tears and crying and weeping. And then they would show the Nazis. It was just the same here. But there wasn't an enemy, was there? Yes...there was an invisible enemy. But I didn't understand it then.

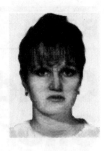

Irina Prokopenko, ninth class, middle school number 2, Harodka

IN THE FIRST DAYS after the accident, we were light-hearted and trusting, we inhabitants of the contamination zone. We lived the same lives as before; children played out in the radioactive rain, we ate pies off open stalls, went to the woods, the grown-ups worked in the fields.

I remember that my parents did not take me and my brother to the May Day parade. They felt a parental concern. But no-one warned us about the radioactive rain.

It was on the Sunday. I wanted to plant flowers round our house. And then it started to rain, and that pleased me, because flowers grow better if you plant them and transplant them when it's raining. My brother ran out to me. We got soaked to the skin, but nevertheless, we got the flowers planted. When we went indoors, our clothes and shoes were covered with a greenish deposit. My brother explained that the wind and the rain had brought pollen from plants, but we know now that this was not pollen, but the terrible dust and ash of Chernobyl...

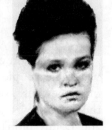

Now I am 17, and for seven years I have been living with thyroid disease....

Natalla Yarmolenka, eleventh class, Brahin middle school

I AM in the ninth class now. I study with different pupils and every year I get more accustomed to our new home... But Chernobyl doesn't leave us alone even here. Last year, one of my new class-mates, my friend,

Maya Kashayed died. The whole school gave her a send-off like a bride. We all stood outside the schoolhouse, and the headmistress rang the bell for Maya, for the end of the last lesson of her short life. A year has gone by since she died, but Maya is still with us. Her place is still there in the classroom. At All Souls and on her birthday, the class makes a wreath, and we go to visit her...
Yelena Kulazhenko tenth class, Hlinitsa middle school, Mazyr County

CHERNOBYL CALLED my Dad, too; he worked for the department of internal affairs. I was in the first grade then, and I didn't understand why Dad was so long away. But Chief of Police Hancharou Alaksandr Mikhaylavich 'was carrying out his service duties in the Khoynitski Regional Department of Internal Affairs in the 30 kilometre zone'. That's what it said on the papers they gave Dad.

I remember how he came home with a voice that was somehow strange and dry. He drank a lot of mineral water; that's what the doctors advised. He told us about the empty villages of Palessie, the domestic animals, howling crazily in the roads and woods, about some marauders...

Two years went by, once again Dad left us. This time he went to Minsk, to the hospital. More and more our house echoed with the word 'Chernobyl'.

CHILDREN OF CHERNOBYL

In 1989, they gave Dad a terrible diagnosis: cancer of the spinal marrow. The doctors in Minsk refused to operate.

Dad went with his sisters to Moscow. They tried to get into a hospital. But the people in the [Soviet] capital were very strange. 'Why have you come to us? We don't have to treat the whole Soviet Union!'...

Then my aunts decided to look for help abroad. It was a long hard road to the USA. Dad was two months in a hospital in Michigan...

And then the long-awaited reunion. Dad came back happy, full of hope. He could walk better; he thought he would be quite well by spring. Enchanted, we listened to his tales about the strange, but sympathetic and kind people he had met in that far-off land.

The American professor wrote a letter to a famous Moscow professor asking him to continue Dad's treatment. But the famous genius said: 'What do you need from me? You've already been here once.'

So Dad was left without follow-up treatment, He began to get worse.

February 1993. Hospital once more. As well as his main illness — pneumonia... The medics were beside his bed day and night. He fought the illness for almost two month, but... on 28 March, Dad died...

That's how Chernobyl took away my Dad. And it took away my birthday too; Dad died exactly on the day I reached 14...

Volha Hancarova, *ninth class, middle school number 18, Barysau*

WE'RE HAVING a physics lesson. The subject is 'Biological action of radioactive radiation'. The words resound: 'radiation', 'irradiation dose', 'isotopes', 'Curies', 'Caesium', 'strontium', 'radiation distress zone', 'Chernobyl'...

I want to stress that I know all these words already, have known them a long long time... Chernobyl has spread its dark wing over every one of us. Here in the Mahileu region, there is no small nook where this trouble has not struck. Klichau county is considered to be relatively clean, but only relatively. Here, too, there are radiation hot-spots. It seems to me that

> *Chernobyl is in my blood,*
> *In both senses of the word.*

At the beginning of 1991 a team of doctors came to our school from the Ministry of Health. They did a check-up on all the children. Many of our pupils had enlarged thyroids. I had this illness too, I went for treatment to the Klichau Polyclinic. And in the summer, I got a trip from the 'Children of Chernobyl' fund to the 'Arlana' youth camp. I spent 40 days there, resting with boys and girls from Klichau county and other parts of Belarus. Everyone felt well and happy. But my heart chokes with grief, when I think about the trouble Chernobyl has brought to our country, about the people who are suffering from it. What is waiting for us in the future? For us, today's 15- and 16-year-olds? And my soul cries out, cries out with grief and despair. Everything possible must be done so that life on this soil is not over for ever, so that towns and villages come to life once more in the radiation zone.

...That's what came into my head during the physics lesson.

Kiryla Kryvanos, *tenth class, Nasieta middle school, Klichau county* ❏

Translated by Vera Rich

Belarusian Popular Front rally, May 1995: an electoral wipe out

Nation in search of a history

Economic chaos, empty bellies and nostalgia for the good old Soviet system are undermining Belarus's advances in democracy. A poor sense of national identity and ignorance of its own past, suppressed and distorted by Soviet historians, are playing into the hands of a repressive government that wants to return the country to Moscow

VERA RICH

1991 and all that

THE REPUBLIC OF BELARUS, an ex-Soviet state of some 10.2 million inhabitants, did not so much win independence as have it thrust upon it. In August 1991, its hardline Communist leaders openly gave their backing to the anti-Gorbachev Moscow coup. When the coup collapsed, the Belarusian hardliners in the Supreme Soviet, fearing the wrath of Gorbachev, made common cause with the small group of pro-democracy People's Deputies and, on 25 August 1991, proclaimed independence.

Belarus had, on paper, been 'independent' for more than 40 years; like Ukraine, it was a founder member of the United Nations and a member of UN agencies, including UNESCO and the International Atomic Energy Agency (IAEA). But this ploy by Stalin to get two extra votes in the UN, in spite of Soviet lip-service to the cultural and linguistic rights of the non-Russian nationalities that comprised almost half the population of the USSR, provided no defence against the long-term aim of *sliyanie* — the 'alloying' of more than 100 ethnic identities into a single, Russophone, Soviet 'nation'. And, though the decision was not at the time made public, Belarus was chosen by the Soviet ideologues as the test-bed of this policy.

It was, in many ways, an appropriate guinea-pig. Belarus had already been subjected to intensive Russification during the nineteenth century, its language forbidden and its mainstream religion, the Eastern-rite Catholic Church (which might otherwise have served, as it did in western Ukraine, as a guardian of national culture) forcibly merged with the Russian Orthodox Church. Following the national revival after 1905, and the window of 'Belarusianisation' in the 1920s, the intellectual elite of Belarus was virtually wiped out (either shot, or terrorised into silence) during the 1930s. World War II meant the loss of one in four of the population — including the destruction of the centuries-old Jewish community which, as its own members eloquently attested, had lived in amicable symbiosis with its Christian, Slav co-habitants for centuries. Furthermore, Belarus in the Cold War era was the most highly militarised area of the Soviet Union, being viewed as the advance post against the NATO powers. And, since Soviet military policy virtually never allowed its soldiers to serve in their home republics, and Russian was the language of the army, the military presence was inevitably a powerful tool of Russification.

Sliyanie worked well. By the early 1980s, there was not a single Belarusian-language school in the capital, Minsk. And since all teacher training was Russophone, even schools in remote rural areas where the language survived were, by force of circumstance, gradually Russified as elderly Belarusian-speaking teachers retired.

Since there was no other obvious unifying shibboleth and symbol of identity — as, for example, the Catholic Church was in neighbouring Lithuania — the language issue took on a special importance to those who wished to preserve national identity. The handful of *samizdat* and expressions of dissent in Belarus during the Brezhnev era concentrated on saving the language. The first manifestation of *glasnost* in Belarus, in December 1986, was a 'Letter to Gorbachev' signed by 28 prominent intellectuals, demanding the linguistic and cultural rights enshrined in the Soviet constitution.

But language alone could not kick-start an opposition movement. That

Dateline Belarus: up, down and out

978 *Rape of Rahnieda: Polatsak and Kyiv at loggerheads*
988 *Conversion to Christianity*
1067 *First mention of Minsk*
1238 *Rise of Grand Duchy of Lithuania-Rus*
1368 *Union of crowns of GDLR and Poland*
1569 *Polish-Lithuanian Commonwealth; end of Belarus as national entity*
1596 *Union of Brest establishes eastern rite Catholic Church*
1696 *Belarusian language replaced by Polish*
1794 *Belarus incorporated in Czarist empire*
1839 *Abolition of Catholic Church*
1905/6 *Czar grants constitution; Belarusian press and publishing allowed*
1918 *Independent Belarusian Republic*
1919 *Belarusian Socialist Republic absorbed into USSR*

was accomplished by two major revelations of the late 1980s: the excavations carried out at the Kurapaty (Windflower Hill) picnic ground in 1988 by a then-unknown archaeologist named Zianon Pazniak, which disinterred the remains of Stalin's victims (some of whose personal effects could still be identified by surviving relatives) thought to be as many as 200,000; and the even more traumatic disclosure, in February 1989, of the true extent of the fallout from the nuclear power station disaster at Chernobyl, in Ukraine, in April 1986. The hitherto-secret maps and data, made public by the efforts of a nuclear physics professor, Dr Stanislau Shushkievich, revealed that more than 20 per cent of the territory of Belarus had been seriously contaminated (p95). Yet, for almost three years, in much of the affected area, agricultural production had continued and no special safety provisions made for the population. Furthermore, two 'hot spots' were revealed far from the main contaminated area, where rain had chanced to fall just as the radioactive plume was blowing back towards Moscow. Chanced? Or, as the rumours now began, deliberately 'seeded to bring down the cloud and save the Soviet capital?' The Soviet authorities denied the allegations, but many scientists studying the fallout patterns remain open-minded.

The shock of these revelations triggered various citizens' movements under the umbrella of what was originally called the Belarusian Popular Front for Perestroika-Renaissance (BNF). In the first-ever multi-party elections to the Supreme Soviet of Belarus in March 1990, several of these movements sponsored pro-democracy and pro-independence candidates of whom 38 were elected to the 360-seat assembly. As a sop to the democrats, Shushkievich was appointed deputy speaker — the one moderate in an otherwise hardline establishment.

A few token pieces of pro-Belarusian legislation were passed: Belarusian was made the state language, and on 27 July 1990, the sovereignty of Belarus was proclaimed. A few days later, on 6 August, (the anniversary of the publication of Skaryna's *Psalter* (p119), a group of young people, under the formal auspices of the Belarusian Language Society, proclaimed the re-establishment of the long-outlawed Belarusian Eastern-rite Catholic Church — an act that challenged not only the hardline rulers of Belarus, but also the Vatican bureaucracy who feared this initiative would rock the carefully trimmed boat of papal *Ostpolitik*.

With independence, the Soviet emblems were replaced by the white-red-white flag and *Pahonia* (Pursuing Knight), symbols dating back to the

Grand Duchy of Lithuania-Rus. Schools and universities hastily introduced courses in Belarusian history. The hardline speaker of Parliament resigned, and Shushkievich took his place, becoming, in the absence of a president, simultaneously the head of state. Hitherto semi-official newsletters became fully-fledged, legally registered journals. Minority religions — including the Eastern-rite Catholics — were granted legal status. The restitution of Church property confiscated by the Soviets began. And a whole sheaf of nation-building initiatives came into being, including a search launched by Foreign Minister Piotr Krauchanka, for a twelfth-century treasure of deep symbolic importance

29 October 1989: on the road to Kurapaty

— the Cross of St Euphrosyne of Polatsak with its double-barred form, as in the symbol of the pagan god, Yaryla (p141) — that had gone missing during World War II. Failing the recovery of the original cross, Krauchanka said, Belarus would use some of its tiny gold reserves, mainly recycled scrap from military electronics, to create a replica of the Euphrosyne cross as a national treasure for the future.

Economically, however, all was not well with the new state. The aftermath of Chernobyl ate up 15 per cent of GDP; Russia's oil and gas producers began to demand what they claimed to be 'world prices' for supplying Belarus's energy needs; 70 per cent of Belarusian industry had,

Once upon a time

FOR GOVERNMENT Pazniak rides not a-wooing,
A harder row to hoe Zianon did pick.
They've drained the marshes, though
   (a deed worth rueing),
So how can he drive Reds down to Old Nick?

Ryhor Baradulin

From Asoby Roznay Proby *(Persons of Different Experience), a collection of political lampoons, written when such commentaries on those in authority were common currency. Translated by Vera Rich*

in Soviet times, been military-related — and now the arms race was over. Ex-hardliners in government and Parliament blocked moves towards privatisation of industry and agriculture and, by their obstructive attitudes, deflected would-be foreign investors to more receptive business climates, usually to neighbouring Baltic states with a similar industrial base.

The first major blow to democracy came in December 1993. Public opinion, which had little understanding of economics, attributed the ever-rising prices and falling standards of living to corruption in high places. Alaksandr Lukashenka, an ex-hardline Communist member of Parliament, levelled charges of corruption against Shushkievich. Although the allegations were unfounded, Shushkievich (who had collapsed in the Parliament chamber with a heart attack) resigned. A few weeks later, when the presidential election campaign began, Lukashenka declared his candidacy, campaigning on a populist programme of an economic upturn and end to corruption — though with no policy proposals for achieving this. But with no expertise in assessing the claims of rival candidates and thinking anything better than the stagnation of the last three years, the Belarusian electorate chose Lukashenka.

In the 18 months of his presidency, Lukashenka, a former state farm boss, has failed to deliver the promised economic improvements. The most disadvantaged strata of society are worse off: he has cancelled even such concessions as the rights of pensioners to travel free on municipal transport; there has been no real movement on privatisation of either industry or agriculture; and major foreign companies who were considering joint ventures with Belarus are now pulling out. Lukashenka puts his hope for economic improvements in close ties with Russia, but the much-lauded customs union with Russia has brought no real benefits to Belarus and the Russian democratic politicians and pro-market economists have little use for Lukashenka's proposals. Lukashenka bolstered his pro-Russian policy with a referendum in May 1995 proposing closer economic ties. No media coverage was allowed to any viewpoint but his own. Though he won approval on the Russian front, the same referendum failed to endorse his wish to change the state flag and coat of arms back to those of Soviet times (minus the hammer and sickle) but the president simply announced to the world that the vote on the *symbolica* had gone his way.

The failure of the long-overdue parliamentary elections in May 1995 (there was insufficient turnout in 141 constituencies) allowed him six months of rule by decree. The country was peppered with directives which included a list of cadres not allowed to leave the country without his written consent (among others, university administrators and the editors of major newspapers) and a blanket ban on all textbooks in the humanities published since 1992. Many of his fiats have been ruled illegal by the Constitutional Court; but Lukashenka does not recognise its competence. He operates by a simple syllogism: the Constitution says Belarus is a presidential republic, he is the president, therefore any decision he takes is constitutional and anyone who opposes him is in breach of the Constitution.

In the past 18 months he has ordered the disconnection of the live broadcasting equipment from the Parliament chamber, sent riot police to arrest democratic members of Parliament, ordered the replacement of outspoken editors and tried to put the few newspapers which dared criticise him out of business (p128). He has banned the newly formed unofficial union, arrested the leaders of the metro strike in August 1995 and sent its supporters to pick the potato harvest. While most of his decrees have been published in the official press, there is increasing evidence of a parallel system of unpublished diktats, delivered as telephoned 'hints' from the presidential aides.

Not surprisingly, the prevailing mood in Belarus is dark. The more politically aware intellectuals and young people are trying to keep alive the message of democracy; Belarusian PEN strives to foster at least the notion of freedom of speech and the printed word. And one positive spin-off from Lukashenka's one-man rule has been the drawing together of Russophone and Belarusian-speaking intellectuals. The notorious May 1995 referendum gave the Russian language equal status with Belarusian, effectively ending the post-independence programmes of positive discrimination. But if Lukashenka hoped this would split the country on language lines and marginalise the 'linguistic patriots', he was mistaken. Resentment of Lukashenka's methods is no less among Russophone democrats than among the most ardent advocates of the Belarusian language. And the most 'patriotic' Belarusian-language newspaper, the bi-weekly *Svaboda* (Freedom) now publishes a regular page of political commentary in Russian.

According to a political joke of the area, when the tanks are approaching, the Poles charge them with cavalry, the Russians attack them bare-handed and the Belarusians dig fox-holes and let the tanks roll safely over their heads. But even the greatest optimist must foresee a long, hard haul for Belarus towards democracy. On 10 December 1995, at the fourth attempt, the Belarusian electorate returned a quorate Parliament, 191 out of a possible 260 seats, but the pro-independence BNF was wiped out. A few pro-democracy candidates, including former parliament speaker and head of state Stanislaus Shushkievich have got through and could form the nucleus of a possible opposition to Lukashenka. But the conduct of the elections underlines the bizarre nature of Belarusian 'democracy'. When Parliament's speaker, Miackyslau Hryb, wanted to address the nation, stressing the importance of a turnout adequate for a quorate Parliament, he was not given access to national television. Russian TV agreed to give him a slot, but President Lukashenka suspended the relay of Russian TV to Belarus 'on technical grounds.' Eventually, Hryb was able to address the electorate — but only via Radio Liberty from Prague.

In the meantime, one may recall the words of *Svaboda* after the trade union association of Papua New Guinea sent a telegram protesting against the suppression of the independent Belarusian trade unions: 'If only we Belarusians knew as much about what is going on in Belarus as the citizens of Papua New Guinea do.' ❏

Vera Rich specialises in Belarusian and Ukrainian affairs. She was the translator of Like Water Like Fire *(1971), the first anthology of Belarusian poetry in English. Sponsored by UNESCO, the book was subsequently withdrawn under pressure from the Soviet censorship*

Frantsisk Skaryna

UNLIKE most post-Communist states, Belarus saw almost no changes in Soviet toponyms. Minsk still has its Lenin, Marx, Kirov, Komsomol and Revolution Streets. One of the handful of name changes, back in the heady days of December 1991, was that of the broad, 15-mile-long highway from central Minsk to the new airport. Originally named Stalin Prospect and renamed Lenin Prospect under Khrushchev, it now became Frantsisk Skaryna Prospect, in honour of one of the greatest cultural heroes of Belarus, the father of Belarusian printing.

As so often with Belarusian history, there is much that remains a mystery about Skaryna. On the evidence of the acrostic prefaces to his publications, he came from Polatsak (Polotsk) in northern Belarus, and considered himself to be 'Lituanus' that is, an ethnic-Slav citizen of the Grand Duchy of Lithuania-Rus. He was a graduate of the universities of Cracow and Padua, the latter degree being in medicine, a qualification of which he seems to have been particularly proud, stressing the title in the acrostics, and showing himself in his self-portrait in the seated pose proper (in medieval and renaissance symbolism) to a physician. We have the various books of the Bible he translated and the devotional works he compiled, in what was then the contemporary language of the Belarusian lands, and know to the very day (6 August 1517) when the first of them, the *Psalter*, appeared.

But virtually all else is doubtful, even his dates of birth and death (?1490-?1552 is the best estimate). Even his baptismal name has been queried: Frantsisk (Francis) is undoubtedly the form he himself used, though in the Stalin era, as 'Georgiy Skorina', he was enshrined in the official Soviet pantheon as one of the *Russian* precursors of Soviet culture!

But other mysteries remain. Did Skaryna meet Luther in Wittenberg? Did he contribute to the drafting of the first Statute of the Grand Duchy of Lithuania? Was he involved, and how deeply, in the secret diplomacy of the time? What do the esoteric devices which appear in his self-portrait and in the decorated initial letters in his publications signify? Was he a cabbalist? An alchemist?

Whatever the answers, his name, and his symbol of the sun and crescent moon, have been adopted by two of the new spearhead organisations for Belarusian culture and language: the Frantsisk Skaryna Belarusian Language Society and the Frantsisk Skaryna Centre for National Education. *VR*

YUYA HURTAVENKO & T I ULEVICH

Saints and heroes

TODAY WE ARE living in a new spiritual climate: Belarusians are becoming aware of themselves as a great nation. But without knowledge of one's roots, one's own history and culture, it is impossible to create a sovereign state.

Our ancestors strove for the independence of Belarus through the course of centuries. But memory is not everything. One has to understand one's history correctly. At different times, many events and the significance of individual personalities in history have been treated incorrectly, often purely subjectively. Hence, in this collection*, the authors try to show the complex path of Belarus through the lives of its great historic figures, starting with [the twelfth-century saint] Euphrosyne of Polatsak, up to the poet Maksim Bahdanovich [1891-1917]...

Out of the Darkness into the Light

The process of restoration is neither easy nor simple. Our history must not seem something remote and unclear. Its task is to inspire our life. It is the duty of those now alive to bring back from oblivion those national traditions and cultural values that could not be preserved in the suffocating atmosphere of Stalinism and [the Brezhnev era of] stagnation. It is this restoration of culture and of the founders of that culture that will determine the fate of the nation.

* © *From the introduction to* Out of the Darkness into the Light: Shining Personalities of Belarus, a Collection of Belarusian Historic Figures for the General Reader, *compiled by Yuya Hurtavenko and T I Ulevich (Minsk 1994)*

Translated by Vera Rich

ANDREW WILSON

In search of a history

The challenge of nation-building is more acute in Belarus than in any of the ex-Soviet republics. For want of a sense of identity, a common history and future destiny, the nation's independence could be lost

FEW IN THE WEST had heard of Belarus before 1991, even under its imperial-Soviet name of 'White Russia'. Now it seems that it may be about to tread a unique path and become the first post-colonial state ever voluntarily to apply for reunion with the former metropolitan power. The leading light in this process is Belarusian president Alaksandr Lukashenka, who won a massive 80 per cent of the vote in the summer 1994 elections by the simple promise that a speedy return to the Russian orbit would solve all the country's political and economic woes. Since his election, Lukashenka has continued to press forward with his plans for economic and even political reunion with Russia. Ukraine has greeted his plans with polite scepticism. In Russia, however, though democratic opinion is increasingly wary of Lukashenka's ambition, a bill to organise a corresponding referendum on some kind of federal union is currently working its way through the State Duma. Significantly, one opinion poll indicated that 30 per cent of Russians would vote for Lukashenka if he were on the ballot in next June's presidential election in Russia.

WHY DOES Belarus seem to place so little value on the independence it won in 1991? Unlike other post-Soviet states such as Moldova or Georgia, which have been plagued by violent ethnic strife since 1991, the problem does not seem to lie with ethnicity. A healthy majority of 78 per cent of the population of Belarus were ethnic Belarusians according to the last Soviet census in 1989, and there are no 'autonomous republics' or non-Belarusian territorial enclaves that might want to break away from Minsk. Only 13 per cent of the population is actually Russian. Nor does the problem seem to lie with regional tensions. There is no equivalent in Belarus of the conflict between east and west Ukraine or north and south Kazakhstan. Belarus is a relatively homogeneous nation-state. Like west Ukraine, west Belarus was part of inter-war Poland, not the USSR, but it remained a rural backwater and never developed the strong local traditions and powerful nationalist movement that mark out west Ukraine from the rest of the country.

Rather the problem seems to lie with Belarusian history. Belarusians are well aware of their existence as a separate nation, but without a well-developed set of myths (used here to denote the role of historical ideas in forming and reproducing national identity), of national history and national character, have little sense of what it means to be Belarusian beyond their collective name and elements of national folklore. Belarusians are aware that they are neither Russian nor Polish, but lack a plausible myth of national descent around which to build a more positive sense of national identity.

To be precise, the problem is not that such a version of Belarusian history does not exist. Belarusian historians, such as Vatslau Lastouski before the 1917 revolution and Usievalad Ihnatouski in the 1920s (president of the Belarusian Academy of Sciences in the 1920s and commissar for national education from 1921 to 1926; Stalin's rise to power led to his fall from grace and he committed suicide in 1931) attempted to construct a complete national mythology including narratives of separate origin and development, heroic struggle against Russian and Polish perfidy, and myths of military glory and cultural achievement, all of which are being refined in the present-day. However, nationalist historians have had severe problems both in constructing a plausible version of the past and in popularising it amongst the Belarusian public.

The first problem is the lack of any real statehood in the past, a 'Golden Age' which contemporary Belarus could be presented as reincarnating. Ukrainians, for example, can claim the early medieval principality of Kievan Rus as their own, Georgians the kingdom of Kartli-Kakhetia that survived until 1801, Uzbeks the empire of Tamerlane. The best nationalist Belarusian historians can do is play up Belarusian autonomy *within* given polities or reinvent periods and places previously ascribed to other national mythologies as in fact *ersatz* Belarusian states. Therefore Belarusian historians deny that Kievan Rus was ever a monolithic kingdom, and assert that the northeastern Marches (the 'city states' of Polatsak and Novahradak) were in fact embryonic Belarusian states in more or less constant warfare with Kyiv. The Lithuanian kingdom of 1238 to 1569 is now described as the 'Lithuanian-Belarusian' kingdom, founded by Novahradak, whose court language and culture were essentially Belarusian. In a particularly fascinating twist, it is claimed that the 'Litviny' after whom the state was named were in fact Belarusians (the Lithuanians were then known as 'Ahamoity', or 'Zmudziny' after the Polish). The defeat of the Teutonic Knights at the Battle of Grunwald in 1410 was therefore a Belarusian rather than a Lithuanian victory. Even the Polish Commonwealth, established via merger with Lithuania (or Lithuania-Belarus) at the 1569 Union of Lublin, is depicted as a decentralised 'dual kingdom' in which Belarusian lands enjoyed almost complete autonomy.

Supposedly therefore the remnants of Belarusian 'statehood' were only finally extinguished in 1793-5, when the final Partitions of 'Poland' brought nearly all Belarusian territory forcibly under Russian control. Moreover, nationalist historians play up the significance of events such as the 1863 uprising (often interpreted in the West as purely a Polish revolt) and the attempt to establish a 'Belarusian People's Republic' in 1918 in order to refute the myth of Belarus' voluntary incorporation into, and ultimate dissolution in, the Russian cultural and political space. National independence in 1991 did not therefore arrive out of the blue, but was only the culmination of a long period of heroic struggle against foreign rule.

In some ways this is not unpromising material. A lineal version of Belarusian history can be constructed with surprising ease by joining up the periods reclaimed from other historiographical traditions (Polatsak-Lithuania-Poland-Russia). However, there is no actual statehood to celebrate; Belarusians have lost (or never enjoyed) other crucial markers of national difference. Local elites were nearly all Polonised after 1569 or Russified after 1793-5. Unlike the Ukrainians the Belarusians lacked

Cossack rebels to stand in their place. Peasant consciousness on the other hand was largely parochial and religious, and increasingly squeezed between the identification of Catholicism with Eastern-rite Poland and Orthodoxy with Russia. Modern historians claim that the Catholic Church provided a means of finding a niche between the two and preserving a separate Belarusian identity (around 75 per cent of Belarusians, it is claimed, were still Uniate in 1793-5), but the Church was forcibly dissolved in 1839 and its modern revival has been feeble.

Moreover, the appearance of lineal continuity in the nationalist version of history is deceptive, as it in fact attempts to piece together a shifting pattern of territories. In contrast to Kyiv's long-standing role as Ukraine's pre-eminent urban centre, Belarus has never possessed a stable capital acting as a single centre of national consolidation. During the Kievan period the main Belarusian towns, Polatsak and Novahradak, were in the northeast; under the Tsars Vilnius (Vilnia) was the main centre of the national movement. Napoleon (and the Germans in this century) would have placed the peripatetic capital at Mahilou (Mogilev). Minsk only became the capital in the twentieth century. Nor has any particular region been able to play the role filled by west Ukraine as the 'Piedmont' of the national movement. The closest equivalent, the area around Vilnius that led the limited national revival movement around the turn of the century, was lost first to Poland in 1921 and then to Lithuania in 1945. The other main town in west Belarus, Bielastok (Bialystok), remained in Poland after 1945.

One last factor is that Belarus does not possess a diaspora community as influential as that of other states such as Ukraine and Estonia, capable of preserving and developing national historiography during the years of Soviet rule. Without much access to alternative ideas, many Belarusians have therefore been extremely vulnerable to socialisation into the Russophile version of local history, which has promoted an entirely different set of myths for most of the last two centuries. In the Russophile version Belarusian and Russian history seamlessly merged into one another (during the imperial and Soviet periods Belarus was referred to as Russia's 'northwestern province' or Belo*russia*, that is linking the Belarusian myth of descent to Russia 'proper' rather than to Rus), and the myth of separate development was displaced by narratives of voluntary incorporation, common enterprise in empire-building and collective struggle against third parties, Germans and Poles in particular. The wars of

'My salvation is in my own hands. In whose hands am I?'

1558-83 and 1654-67, and the rebellion of 1794, during which nationalists claim that millions of Belarusians died at Russian hands, are more familiar to most Belarusians as periods of peaceful incorporation during which the local 'Orthodox' were saved from the Poles.

Soviet historiography even denied the very possibility of a separate Belarusian history by asserting that Belarusians and Russians (and Ukrainians) were descended from a single 'old Rus nation', that is the Orthodox population of Rus. The Russian annexation of Belarusian lands in 1793-5 was therefore 'reunion' rather than (re)conquest. While the history of Estonia or Georgia was subject to mere misrepresentation and distortion, the Belarusian past was simply wished away. (Belarusian nationalists by contrast argue that the eastern Slavs were always divided; intermingling between local Balts and the Kryvychy tribe who founded the city of Polatsak created the foundations of the Belarusian ethnic group, whereas the Russians are supposedly descended from a mixture of Finno-Ugric and more northerly Slavic tribes such as the Sloviany and Viatychi).

Given sufficient time, future generations might be socialised away from the Russian/Soviet version of national history. All national histories are initially delicate constructs and the nationalist version of the Belarusian past is so contrary to the Russophile version that it is hardly likely to be accepted overnight. Nevertheless, if Belarus survives as an independent state for long enough, then the mere fact of its separate existence may gradually help to verify the nationalist version of history *ex post facto*. However, at this crucial conjuncture Belarusian nationalists are unlikely to be granted control of the school system to facilitate this task. In fact movement is in the other direction. In August 1995, President Lukashenka announced the formation of a shadowy committee with a vague remit to 're-examine' textbooks written since 1992 and to reassert the traditional predominance of the Russophile version of Belarusian history.

Why does all this matter? The single most important factor in building a putative nation-state is neither economic strength nor military might, but a nation's sense of its own identity, common history and future destiny. In order simply to function as a normal member of the international concord of states then a would-be 'nation' must at the very least conceive of itself as an independent subject. Many Belarusians do not. This is not to say that Belarus will not survive in some form or other. Having won independence in highly unusual circumstances in 1991, Belarus' statehood is now underpinned by the powerful inertia of the international system and the self-interest of Belarusian elites in remaining big fish in a small pond. Moreover, Russia itself has to consider the economic and political consequences of welcoming Belarus back into the fold. Nevertheless, the politics of identity often trumps the politics of pragmatism. If Russia obtains new rulers in 1996 it may begin to take Lukashenka's overtures more seriously. ❑

Andrew Wilson is a member of the Post-Soviet States in Transition Programme at Sidney Sussex College, Cambridge, UK

President, go home!

THEY look at the red sky
and hope it will be raining.
They look at the dry soil
and hope it will grow green.
And we reserved the hope for us
but it didn't sprout.
Yet we reserved three words for us:
president, go home.

THEY wake up and get their beer.
It seems to them everything is OK.
They're sniffing out: 'What? Where?
    How much?'
and hope it's gonna be cheaper.
And we reserved the freedom for us
but you can't drink it like beer.

YET we reserved three words for us:
president, go home.
They don't know the word 'I'.
They don't know the word 'Word'.
They don't know anything about love.
I'm not even using this word.
And we reserved the word for us
but they won't hear it.
Yet we reserved three words for us:
president, go home.

Kasya Kamotskaya and her rock group Novaya Neba (The New Heaven) have been banned from public performances and on TV and radio

Minsk, Pinsk and other places

'The Jews who live here, in their new homeland, have taken over more from the Belarusians than the Belarusians have taken from them. The mighty force of the Belarusian land has given a special spiritual and physical appearance to the Belarusian Jews. Now they differ from all other Jews, and throughout the whole world they are called "Litvaks".' The Jews of Belarusia *Zmitrok Biadula (Minsk 1918)*

THE 'DIFFERENCE' Biadula notes was seen most clearly in the cities of Belarusia where Jews were at the heart of cultural and economic life. Unlike their counterparts in Poland, Lithuania and Ukraine, where Jews were driven to carving out an existence on the fringes of society in the harsh confines of the *shtetl*, Belarusian Jewry was at home on the boulevards of Minsk, Pinsk and Vitebsk, where, contemporaries testify, they lived in harmony and mutual dependence with the native population. Jewish intellectuals were instrumental in the Belarusian national revival of 1906.

By 1897, four centuries after the first Jewish merchants arrived in Minsk, 13.6 per cent of the Belarusian population was Jewish. In the cities they were a majority: 52.3 per cent in Minsk out of a total population of 91,500, 52.4 in Vitebsk and 72.4 in Pinsk.

Tales of the good life in Belarusia, as well as the growing fame of the Minsk *Volzin yeshiva*, travelled abroad. Young Jews from Germany and the USA came to study; older immigrants fled pogroms elsewhere.

Only a generation later, Stalinist purges and Nazi occupation had all but wiped out Belarusian Jewry. By 1941 it had been halved; between 1941 and 1945, Nazi genocide destroyed hundreds of thousands more. Of a population of 1.5 million in its heyday, Belarusian Jews today number only 73,000.

While Soviet historians persistently denied the Holocaust on Soviet territories, alone in the USSR, Belarusian writers put the truth on record in their fiction.

In the years following World War II, remaining traces of the Jewish presence were systematically erased. By the late-1960s, only one of the 100 or more synagogues and prayer halls in Minsk remained; the Jewish theatre was abandoned; their centuries old cemetery ploughed up for a hotel. Despite this, the Jewish community is once again playing an integral part in building the intellectual and artistic life of independent Belarus.

Anna Feldman

ALEXANDER BELY

Freedom of oppression

'Go ahead, give it a turn! I work better under pressure!'

With its papers banned and no access to state-owned TV and radio, the opposition has no way of getting its message across

FOR THE LAST few days, I have not received the newspapers I regularly subscribe to. The post office will no longer deliver them. They are not to be had in the kiosks of Sayuzdruk, the state-owned news agency that holds a monopoly on distribution. All state-owned printing houses (there are no private ones in Belarus) have been instructed to annul the contracts with my papers since our president, or *Bats'ka* (The Father) as he prefers to call himself, has decided I no longer need them.

He has already rid us of a few other redundant items: our national flag and coat of arms, for instance; Parliament (the old one is past its sell-by date but continues to sit regardless; the new one as yet has too few deputies to begin its work); deputies' immunity; free trade unions and so on.

Regardless, these foolish newspapers clutch at straws: they now print in Lithuania and are attempting to create an independent distribution service, forgetting that there are customs officers to get past at the border and that accounts to pay for all this can be frozen at any time. The Father does what he wants with his children with impunity.

There are those who console themselves with the thought that this was how it was in Pinochet's Chile: 'It's only natural: wait and suffer. Dictatorships facilitate the transition to a market economy: no political freedom but the economy booms (and we all get rich).' Maybe.

I'm not sure: maybe there is nothing in common between the human rights that are so important to a few intellectuals and the economic opportunities critical to the 'average' person. Maybe. But here is a typical scene from Belarusian 'business' life: Ivan Titenkov, the head of the president's administrative department, takes a look at a nice building in the centre of Minsk. 'This will do nicely,' he utters contentedly. The next morning you read in the newspapers the president's decree assigning that building to Titenkov's department. Most newspapers explain how beneficial this measure is for the well-being of the ordinary people. A few black sheep, however, immediately start to speculate on Titenkov's objectives. Nor can they understand why the bank of Father's close friend should be saved from bankruptcy by a merger with the biggest state bank. And, of course, their pages are filled with absurd notions on the state and its functioning. It is these papers that I am finally delivered from: *Imya* (Name), *Svaboda* (Freedom), *Narodnaya Volya* (People's Will), *Belorusskay Gazeta* (Belarusian Newspaper), *Belorusskay Delovaya Gazeta* (Belarusian Business Newspaper). Thank you, comrade Zametalin.

There are, of course, six official and semi-official daily newspapers. If you're after something more than reports on the infinite wisdom of our president, the potato harvest and milk yields, they are not a lot of use. Of these six, only *Narodnaya Gazeta* (People's Newspaper), with its pluralist past, tries to observe the proprieties when flattering Father — which does not mean it dares to criticise him. Their total circulation, two-thirds of the circulation of all Belarusian dailies and periodicals, hardly exceeds 800,000 (in a country with a population of 10 million), too little to convey all that is to be had of Father's wisdom. To remedy this, comrade Zametalin decided to restore compulsory 'political information sessions' for all workers.

Oddly enough, retired colonel Vladimir Zametalin headed the election campaign of the former Belarusian leader, Vyachaslav Kebich, during the 1994 presidential elections and, in the process, unleashed the avalanche of mud-slinging directed at Lukashenka, himself denied all access to the media. As a result, Lukashenka promised, if elected, to liberate the Belarusian press for the first time in its history.

However, shortly after his inauguration, Lukashenka appointed his former persecutor head of the department of public and political information, presumably in recognition of his superior propaganda skills. Officially, this department is nothing more than the presidential press office; in practice, its influence penetrates deep into all spheres of public life. It scans every newspaper, book, TV and radio programme, film and theatre performance in search of the slightest sign of sedition. It banned all school textbooks created during the post-independence 'thaw' of 1991-1994, written to teach children the history of their own country, not that of Holy Russia or the Heroic Soviet People. In the event, since there were no textbooks yet prepared in conformity with the official presidential ideology, the ban was suspended perforce. The latter is based on the mystical 'Unity of all Slavs' against the world and denies individual freedoms on the grounds that these are 'alien to Orthodox Slavs'. The prevailing paranoia was illustrated on 12 September 1995 when two US balloonists were shot down as 'spies' although they had permission to cross Belarusian territory.

In late October 1995, comrade Zametalin was sent on a secret mission to Iraq to congratulate Saddam on his convincing victory in the presidential referendum. Lukashenka's minions are hard at work convincing public opinion that 'international sanctions imposed on the peaceful Iraqi people by western imperialists are unjust and should not be complied with.' Some predict that this is the start of a campaign to create the Minsk-Moscow-Baghdad axis — Greater Eurasia. Our small but proud state is also strengthening political ties with Cuba and North Korea, neither of which would be happy to have a free press operating in the country.

We are now in the third phase of the Father's promised development of a free press. The first was in December 1994 with opposition deputy Siarhiey Antonchyk's exposure of corruption in Father's entourage. Papers that attempted to publish Antonchyk's report appeared with blank columns and editors of government-subsidised newspapers were sacked.

Phase two preceded the May 1995 referendum. Phase three, in which we are now embroiled, was a safeguard in the run-up to the 29 November 1995 parliamentary elections. Lukashenka has made it clear that he would prefer to manage without any Parliament; but with the West in love with representative institutions, and the country desperately in need of foreign investment and expertise, he is ready to put up with it — provided Parliament is elected with no interference from the press.

Yet Lukashenka's attempts to destroy the opposition press have, so far, only strengthened its influence. In spite of persecution, maybe because of it, timid pamphlets are turning into serious and respectable publications. ❑

Alexander Bely graduated from the Belarusian Technical University in 1989

Press edict, 1 September 1995

TO THE EDITORS of newspapers, directors of information agencies and other mass media outlets in the Republic of Belarus.

In the wake of the May referendum and while preparations are under way for the second round of elections to the Supreme Soviet, it is clear that part of the media is not fulfilling its obligations under the law 'Concerning the press and other means of mass information', which defines the responsibilities of editors and journalists, namely the accurate and conscientious delivery of information to the public.

Government leaders and workers, the security services, the control services of the Supreme Soviet, the administration and the President of Belarus are victims of disinformation.

The editors of particular newspapers and radio and TV programmes distort the results of the referendum, the internal and external policies of the government and the spirit and purpose of reforms in the education system designed to ensure social tranquillity in the Republic.

There are several instances where the media gives an inaccurate picture of the activities of structures which are specifically designed to uncover bribery, corruption and the abuse of power by responsible officials, namely the Supervising Board (*Kontrolnaya Palata*), the Supreme Soviet and the Control Services of the President of the Republic.

At the same time, perhaps deliberately, they present a flattering picture of the activities of businesses, private financial groups and political leaders who openly espouse nationalism, Russophobia and deepen the divisions in society.

Objective, truthful and constructive information about life in the Republic, about the working man, high moral standards and the history and future of the Belarusian people is more and more difficult to find in our papers, radio and TV.

Honesty, human decency, duty and responsibility to the people and the fatherland are being replaced by things far from the heart of the people, like petty political squabbles and the so-called 'New Culture'.

Although this has been tolerated by the Ministry of Justice and the Ministry of Culture and Press, a majority of the people have become frustrated and alarmed.

As Head of State, it is my constitutional duty to bring these urgent problems to your attention. For the sake of national stability and the advancement of the Belarusian people, I hope that the media will undertake the necessary changes.

President of the Republic of Belarus, Alaksandr Lukashenka

ZHANA LITVINA

No prospects

ACCORDING to Zhana Litvina, a broadcaster and founding member of the recently formed independent association of journalists, any hope of broadcasting independence was dealt its final blow in the run-up to the May 1995 election and referendum.

The authorities made their first move in December 1994. 'From now on,' said President Lukashenka in a meeting with youth organisations and students, 'we are going to take an active role in influencing journalists and intervening in their work.' Furthermore, he warned, 'I intend to be president for at least 10 years and journalists should get used to the idea.'

Lukashenka was quick to follow up on his promise. On 8 May, *Prospekt*, one of the few TV programmes that could lay claim to a discursive, analytical approach to controversial subjects like language, history or the national symbols — and high in the ratings polls — quietly disappeared from the air. Its director was dismissed while its audience was distracted by the glitz of official junketings in the Minsk Dinamo stadium. Minsk local TV, Channel 8, was taken off the air for 'technical maintenance' in mid-April and the government monopoly of the broadcast media was complete. 'There was no way we could educate electors or act as a forum for debate,' says Litvina. 'The government was determined to subordinate public opinion to its aims.'

By this time, the government was ready for more positive action. Its 'Plan to ensure the proper conduct of the 14 May referendum' was specific and detailed. This 'old-style epistolatory masterpiece' — Litvina's words — set TV and radio journalists a simple task: 'to form among the population the desire to take part in the referendum and to vote "yes" to its questions.' Youth radio was instructed to 'include sociological questions, brief speeches, interviews with actors, youth leaders, organise discussions among young people from various occupations and concerts with titles such as "Both languages are dear to us", "We sing in two languages", "Musicians for two languages" — all interspersed with "short speeches" on carefully selected topics.'

Journalists in state radio were instructed, relates Litvina, to 'prepare a series entitled "Is the Russian language native to Belarusians?" with philologists and historians. Other sexy subjects included "We are all from the Slavonic cradle" and "What language did Skaryna speak?" A "permanent team" of speakers from appropriate institutions such as the Slavonic Council, sympathetic to the president's pan-Slavic, Russophile ambitions, was recommended.'

There was nothing ambivalent in the presidential plan: these were the 'desirable people of culture' who had previously spoken for association with the Russian Federation and for the existence of both languages. 'Society has split into two distinct camps: those who are already celebrating their victory over Belarusian independence, and those who will never forgive the insult to the national symbols and the humiliation of the government's determination to hand the country back to the imperial power.'

Litvina points out that while any chance of developing an independent broadcasting service is out of the question, the government has licensed the operations in Belarus of the interstate Moscow-based TV and radio company Mir. According to the Ministry of Culture and Press's own data, Russian TV is received by twice as many viewers as the national network; Russian radio stations cover 50 per cent more listeners than the state service.

'How', ask Litvina on a despairing note, 'can we build the state, develop a national consciousness, promote economic reform or develop democracy and a civil society without an alternative broadcast media?' ❑

Interview by Judith Vidal-Hall

'How shall we manage without the independent media,' asks the occupant of the lavatory (Belarus)

ADAM GLOBUS

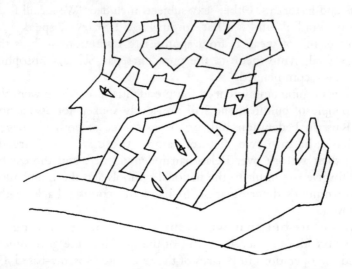

Demonicameron tales

The house demon

THE YOUNG WOMAN was in the house alone.

In the evening, there was a persistent ringing at the door. The woman thought that it was her neighbour, and opened up. On the threshold stood an unknown, bearded man, in a black leather coat, and with a hat pulled down over his eyes. The woman tried to close the door, but her guest stopped her with an impressive gesture of a hand wearing a glittering ring.

'Who are you? What do you want?' the mistress of the empty dwelling whispered almost inaudibly.

'I am the House Demon. Everything in this house belongs to me.

And you, too, belong to me.' The man came into the house. 'Get undressed!'

The woman took off her red robe with the white roses and stood there in her black silk underwear.

'Take off everything?' she asked his pointed shoes.

'Everything!' came the voice from under the dark hat.

The lady of the house took off her underwear.

'Turn your back to me, and bend over.'

The woman, she knew not why, obediently carried out the House Demon's commands. She turned her back to him and bent over. He did not even remove so much as his jacket when he performed what people usually refer to as The Act.

Apart from a cold fear, the woman felt nothing.

When the young woman in her red robe with the white roses was smoking a thin cigarette, she was convinced that the House Demon was the fruit of her fantasy.

The wood demon

At Catholic Easter, a student and a lad from the trade school went into the forest with a girl-student.

The student and the trade-school lad each carried two 700-gramme bottles of 'Radiant' wine, and the girl carried a bundle with bread and slab candy.

Snow was lying in the forest, and they walked for a long time, to a glade where, the trade-school lad assured them, there shouldn't be any snow.

And, in the long-sought glade, there was no snow; it was covered with last summer's grass, dried by the Easter sun.

They drank the first toast to the resurrection of Jesus Christ. The second to the Saviour. The third to spiritual love for one's neighbour. And the fourth to unspiritual love.

While the student was opening up the last bottle, the trade-school lad fell asleep. The two students drank in gulps and helped the cheap wine down with the slab candy.

'Brrr!' He shook his head.

'Ugh. I don't want any more', she whispered.

'Shall we go for a walk?', he suggested.

They walked through the forest, until they arrived at another glade, where, just as in the first, there was no snow.

He blocked her path and asked:

'May I kiss you?'

She agreed.

He kissed her with a very long kiss, and asked:

'Did you like it?'

When she said that she liked it, he asked:

'May I kiss your breasts?'

She blushed as red as a rennet-apple, and said:

'All right, but my breasts are very big, you can't tell under woollies and coats, but they really are so big, it's simply frightful!'

He undid her coat and woolly and kissed those really very big breasts.

Then the two students lay down on the grass warmed by the Easter sun.

'Not like that!', she said.

'Then how?', he asked.

She got up on her knees, bent over and said:

'You come from the back. I like it better that way, because mine is closer to the back!'

He said many caressing and tender words to her, and she said many nice words to him.

Afterwards, they returned to the glade where the trade-school lad still slept, half-poisoned by the rotten wine.

The students drank up the 'Radiant' and ate up the bread and slab candy. She fell asleep. He enjoyed watching how beautifully the girl and the trade-school lad slept, not even angels could compare with them. He enjoyed it, but then he got bored, and went back from the forest to the city.

The trade-school lad woke up and saw the student girl asleep on the grass. He crawled over to her and kissed her. She woke up, but, out of curiosity, pretended she was still asleep. The trade-school lad realised that the student girl was only shamming not to have woken. But out of curiosity, he began to undress her. When he had undressed her completely, he got undressed too, and she pretended that she had just woken up, and said:

'Not like that!'

'But I like it like that', said the trade-school lad, when he heard a strange voice, not a student's voice.

'You can't do it like that!'

He sat up, and she sat up. They sat on the grass and stared at this person who had interrupted their fun.

The interloper had a moustache, and wore a heavy, black trench coat and a bandolier, boots and a flat, peaked cap, with a lacquered band.

'You can't do it like that' he repeated and added, 'I am the Wood Demon. I keep order in the forest. You've been drinking and eating... But who's going to pick up the wrappers from those slabs of candy? First pick them up, and then make love.'

Under the eye of the Wood Demon, the student girl and the trade-school boy picked up the litter.

At Orthodox Easter, the student girl and the trade-school lad celebrated their wedding, but they did not invite the student, because he had not picked up his wrappers.

The *Kaduk*

THE COOK from the school in Rudnia used to make moonshine. She even got fined for it, 400 roubles, no less. And she had to sell her ducks and geese to pay that fine. But even after those difficulties, the cook didn't give up her occupation. She distilled even more, only now she sold it through people she knew or by prior arrangement. You couldn't just turn up, knock at the window and get served.

Just at that time, there were drainage men billeted in the next village. Our marshes are like a sea, wherever you cast your eye. And there are quagmires too, and 'devil's eye' bogholes. Well, you know what drainage men are like! No-one even asks to see your papers for a job like that. Well, they agreed with the cook that in the depth of the night she'd bring them five litres of moonshine — a three-litre jar and a two-litre. They paid her 15 roubles for this. That's how long ago it was!

So one night she put the jars in a shopping bag, packed them round with newspapers and set off to her customers. She was going along the path by the river, when she stumbled and said: 'May the *Kaduk* take you!' Then she went on further, along beside the Vialla, till she saw someone standing on the bridge. And she had to get across to the other side. 'Maybe it's one of the drainage men come to meet me?' she thought, and stepped on to the bridge. And then she saw in front of her, not a man, not a beast, but something terrible, with a great shaggy head, small, pig's eyes, and a wide throat. And this horror was dressed in an ordinary drainage man's gear: cheap, padded jacket, tarpaulin trousers, and gumboots with the tops turned back. The cook wanted to get past this horror, but it blocked her path. She grew frightened, but she put on a brave face and said:

'What are you doing here, man, at such a time?'

And the horror opened its jaws, which were filled with thin, curved teeth like a pike and started to growl. The poor woman turned cold inside and her voice trembled:

'Perhaps you'd like some moonshine, man? I've got some!'

The cook put down her bags on the bridge, took out the two-litre jar and gave it to the horror. It tore off the seal with its teeth, drained the moonshine, threw the jar into the river, and licked its lips as if it had been drinking not moonshine but the weakest bread beer.

'Well, then, perhaps I can go?' asked the cook and made as if to go. She had picked up her bags and turned her back to him, when she felt his hand or paw on her shoulder. And the nails on that paw were sharp, long and as shiny as iron.

'What are you thinking of, man? We don't even know each other!' She looked round at him and saw his open jaws. In which, instead of a tongue, there was a burning flame.

'Well, if you want it that much, then you'd better do it! Only I'll take everything off myself, because you could tear my blouse with those talons.'

She got herself ready, and the horror, too, got itself ready. But when the cook saw the creature's member, she almost fainted, thinking that she couldn't possibly receive such a huge organ into herself — a stallion's was nothing in comparison. But her other dread, that the horror would take her life, drove out the first fear. The cook took the creature's member in her hands, and began to guide it in as best she could. And somehow she managed it, and her fear was so great that she felt no pain.

The woman endured this love more than an hour, and then her strength left her. Her flesh seemed to turn to water and her consciousness became clouded. And a deathly sleep overcame the woman and the beast opened its jaws to the whole width of its throat and swallowed the poor dear up.

For a long time in Rudnia, there were various rumours going round. Some accused the drainage men, saying that they had robbed and raped and killed the cook. Another version was that she had run off with the pay-clerk who brought the drainage men's wages. But one very old

woman said this, just as I've written it down:

'The *Kaduk* took her, the hussy. That creature would like to carry off all people, if it was free to come into this world when it liked. But God has made things so that it has no power over itself, but has to wait until it is called. Don't call the *Kaduk*!' ❑

Adam Globus *is a writer and artist with a strong interest in pushing the limits of censorship and control as far as they will go. Erotica is his latest means of challenging the system. Significantly, his collection* Damavikameron (Demonicameron) *which combines the erotic with the animistic spirits of Belarusian folklore, was published not in Belarus, but across the Polish border in Bialystok, 1995*

Translated by Vera Rich

BOLHA IPATAVA
Crossing the line

IN 1971, Bolha Ipatava wrote a small work on Euphrosyne, a small-time twelfth-century saint, no martyr she, simply a nun who did a good job by way of furthering education and literacy in the principality of Polatsak to the Catholic rite. In the same year, along with four male contemporaries, Ipatava was awarded a coveted prize by the Young Communists which she refused in protest at the rampant Russification of the country under Brezhnev, and became the despair of the national intelligentsia. The influential local Party paper, *Sovietskaya Bielarussya,* savaged her and her work. It was, she recalls, 'the first time anyone had talked about Euphrosyne since the country became part of the Soviet Union. Atheism was tops; religion was out; and anything Belarusian was definitely not on the agenda. Writing about a nun was tantamount to writing a eulogy of the Czars.' And, she adds with the wry wisdom of hindsight, 'I was *very* young.'

In 1986, wiser maybe but with no less determination, Ipatava won the Soviet Order of Merit for being among the first voluntarily to enter the contaminated areas of Belarus where she read her poetry to the victims of Chernobyl.

Between those dates, she was a non-person on the literary scene, unable to publish a word. It didn't stop this gentle, compassionate woman from writing.

And, when the opportunity came with *perestroika*, she was one of the first to plunge into the national debate, passionately arguing, first on TV and then in her own magazine *Kultura*, the need to revive Belarusian language, culture and history, to give back to the people all that had been taken from them.

Inevitably, first her TV programme and then *Kultura* came under attack; publicly silenced on all that is important to her for the second time, disappointed by the performance of the BNF, there is still the poetry.

Interviewed by Judith Vidal-Hall
Unpublished poems translated by Vera Rich

Chernobyl

HOUSE. Hundred-year oak. A well-spring flowing.
And storks winging from a distant strand.
'Human, grant that we may make our home here,
Very sweet to us appears this land.'

Wide the beaks of stork-chicks gape, appealing,
In the low-land fields the years creep by,
Joyfully above the house soar, wheeling,
Two white crosses, like a guard on high.

... Summer rich in sap is growing stronger.
Like molten quick-silver the sun's light.
Two sick children in the nest still linger,
That will never wing away in flight.

'I gave you no warning that the pasture
Is all poisoned to eternity,
Like me, your babes will die of this disaster!'
And the Human cackles evilly.

On the grizzled she-stork as last portent
Of dangers an eternal dream descends:
There a well-spring bubbles with dead water
And beside the house a Werewolf stands.

Trinity (after Andrej Rubliov)

THE ANGELS of our land now sit, forlorn.
Around a bitter cup: wings droop, enfeebled,
Here swords against close kin again are drawn,
Jaryla's cross rejected by the people.

Time, therefore, once again to seek the dust,
Once more accept the earth-bound lot we're given.
Forget the holy for a daily crust,
And live without lifting our eyes to heaven...

VALENTIN TARAS

The tale of Tuteishin

VALENTIN TARAS, Belarusian poet and writer, tells of a time, 25 years ago, when he wrote *The Amazing Adventure of Tuteishin* (*tut* as in 'here', therefore one who has no past nor origins, literally, 'I'm just a person from here'). 'It's the story of a man who has a thirst for the truth, especially the truth about himself. He plagues colleagues, neighbours and bosom friends with the question: "What do you think of me? What sort of guy am I?" People either laughed the thing off or, embarrassed, mumbled some incomprehensible reply.

'One day, while queuing up for beer, Tuteishin picked a fight with an old veteran who was trying to jump the queue. The man gave Tuteishin a hefty blow to the head with his stick. There was a click in Tuteishin's head and, suddenly, he had acquired an amazing gift.

'From that moment, he could hear everything anyone was thinking. A person would look at him in silence but Tuteishin heard as clearly as though the words were spoken exactly what that person was thinking of him. It was never anything to his advantage: "What a mug," one would say. "What a ghastly, slovenly creature." Nor did they mince their thoughts, but enumerated his shortcomings in minute detail. Even his wife had nothing good to think of him.

'What was the result of all this? It never for a moment occurred to Tuteishin that he should mend his ways, repent or improve himself. Instead, he prayed daily that God would deliver him from his amazing gift as fast as possible. He searched high and low for the veteran who had started it all and, having found him, provoked a second fight. Once again, the old soldier dealt him a blow to the head; once again the click; once again, Tuteishin was his old self, minus the gift. He never again bothered his friends and acquaintances with the question: "Tell me, what do you think of me? What sort of guy am I?"'

For obvious reasons, says Taras, the story was never published. But, he adds, it seems as though his little fantasy is being lived out in Belarus

Government election rally, Minsk 1995: ignoring the crimes of the past

today. In the early days of *glasnost*, when only those at the top of the old Soviet structure were exposed and reviled, people rejoiced in their new freedom. 'But,' Taras continues, 'when people started to drill deeper into the monstrous iceberg that was the socialist state, analysing the samples that came up from its depths, it became clear that the entire mass was swarming with the microbes of lies, time-serving parasites, sloth, irresponsibility, greed and xenophobia. The iceberg crumbled and people felt uncomfortable on their little post-totalitarian fragment. Individuals and society as a whole heard the truth and it was bitter and painful. The truth turned out to be a burden that only those who were liberated from the past, who could think freely and independently, could support.' He quotes: 'What is freedom of speech/Without free thought?/An empty purse/A shoe without a horse.

Taras excoriates a government that muzzles the press to keep its people in ignorance of unpleasant truths about the past; equally he has little patience with the people that put it there. 'Society in general demonstrated (by voting for Lukashenka) that it didn't want to know anything about itself that would demand repentance, remorse, any sense of responsibility for the state's crimes and shortcomings.' In short, he reflects, we all want to have a mirror that reflects only our own opinion of ourselves and to smash the one that reflects our true selves. 'It's not only governments that make a wry face, get angry and stamp their feet when they see themselves reflected in an objective looking-glass. Most of the people agree with their "Father" and think these scribblers who criticise their "Father", even caricature him, should be horsewhipped, even condemned to death if they do not respect the people's choice.'

So what's to be done? Should the writers of Belarus bide their time, keep their heads down and wait for the storm to pass? Not at all. 'We have to speak, write the truth but in the knowledge that though truth will not win, it is itself invincible.'

Is there, then, only heroic martyrdom to look forward to? 'We are like the man in Ionescu's *Rhinoceros* who, when all those around him turn into rhinos and surround his house, roaring beneath his window, thinks: "Maybe it's convenient to be a rhino, advantageous, pleasant even. I'd love to become a rhino, but I can't do it. I cannot conquer my nature: I can only be a man." And he takes down his gun...' ❏

Interview by Judith Vidal-Hall

Plaxes from *Imya* by Dmitry Plax

Plax — *a sniveller or cry-baby* — *is the name of one of our most pungent and unshackled writers/cartoonists,' writes Alexander Bely. His 'Child with the ruthless jaws', a common image in children's books, may be the president, or the totalitarian and capricious state that grinds everything to rubble. In the first cartoon 'the child' aims his missiles at the balloons — the Constitutional Court, Parliament and the independent press. In the second, he demands that all banks are united. The jars or bottles in his hands are a pun on bank/bottles, the same word in Belarusian*

GORDON BROTHERSTON

Native testimony in the Americas

Native Americans' tenacity and resilience against overwhelming odds is demonstrated by the poetry of a new literary resurgence

As a result of invasion from beyond its shores, the so-called New World has suffered uniquely: in the course of just a few centuries its original inhabitants, though settled there for millennia and countable in many millions, have come to be perceived as a marginal if not entirely dispensable factor in the continent's destiny. Educational systems in its modern nation-states seldom relate surviving indigenous peoples to their deeper past, and history, like literature, law and philosophy, is most often said to have begun with Columbus. In 1927, the poet César Vallejo noted how western imperialism had robbed China of all but its land and people and wondered whether even those minima would survive in his native Peru.

Despite much triumphalism, the 1992 quincentenary had the advantage of re-focusing attention on these issues, not least among America's native peoples themselves. In July 1990, representatives of over a hundred nations gathered in Quito (Ecuador) at a continental conference called by the Indigenous Alliance of the Americas, to review the experience of the last five centuries. Agreement was reached on eight points, in a declaration that begins: '[We] have never abandoned our struggle against the oppression, discrimination and exploitation imposed on us as a result of Europe invading our ancestral territories.' Going deeper than political alliance, their unanimity drew on notions not only of native dispossession, but of human survival. Unlike the international capitalism that to date has been responsible for such abuse, Articles Three and Six of the Quito Declaration refer to the communally held faith in the earth matrix and to the life-philosophy that explicitly defends natural resources. Note was also taken of how in American 'Third World' states, 'national juridical structures...are the result of...neocolonisation' (Article Eight).

This resurgence has its literary as well as its political edge. Affirming his identity as an Acoma, Simon Ortiz put it like this: 'At times, in the past, it was outright armed struggle, like that of the present-day indians in Central and South America with whom we must identify; urgently, it is often in the legal arena, and it is in the field of literature.' This coincides with the Brazilian Márcio Souza's view of literary engagement as 'counter massacre': here, native language and history themselves serve as a resource in the struggle against the physical and intellectual violence of neo-colonialism.

Chile has recently witnessed a Mapuche literary renaissance led by Sebastian Queupul, Martin Alonqueo, Elicura Chihuailaf, Victorio Pranao (and others who have appeared in the 'Küme dungu' series of texts published in collaboration with the University of Temuco), and, above all, by Leonel Lienlaf, whose collection *Nepey ñi günun piuke* (The bird of my heart awoke) appeared in 1989.

For Lienlaf the mountain that saved people from the flood, Threng-threng, still serves as a promise of refuge when seen from a boat out at sea '*Ül pu challwafe*' (Song in a boat); and in Temuco, south of Santiago and at the heart of Mapuche territory, another mountain, Ñielol, remembers the quite recent times when none of the houses there was western:

Kautinleufü	The river Cautin
ranginmew müley	through the middle
ngümanmew nagküley	runs crying
Temukowariapüle	through Temuco
ngümanmekey	crying
Ñielolwinkul	Ñielol mountain
anüley lelitupelu	sits watching
füchakeruka	large houses
mapuchenoruka	non-Mapuche houses
rakiduamküley	it thinks
Temuko-waria	Temuco town
mi iñchemew	beneath you
umagtumekey	are sleeping

NATIVE AMERICA: POETRY

Mapuche women, Chile

ñifüchake cheyem	my ancestors
Pewmanmew	Dreaming their dream
Müley yengün	they are
ka witrumekey	in the river
leufümew	runs
ñi mollfün	their blood

Yet in '*Chol kin munguey*' (They tore the skin off his back), the wounds inflicted by the savage invasions of the late-nineteenth century, on both sides of the Andes, threaten even the idea of native coherence.

In a further poem '*Rupamum*' (Footsteps), the Spanish vocabulary that denotes the means of oppression (cross, sword) is made to intrude painfully into the Mapuche text, in images of considerable violence:

mutrungreke trekan	Through the tree-trunk
chew ñi rupamum füchake antikuyem	I walked a hundred generations
ngümanmew ayenmew	suffering laughing
dakinmew ñi pewma	within my soul

Quechua musician, Peru

ina pen kine cruz katrünmaetew	then I saw a cross severing
ñi lonko	my head
ka kiñe espada bendecipeetew	a sword blessing me
petu ñi lanon	before my death

Facing invasion today in Peru, in a civil war brought on by centuries of racist outrage against them, the Quechua draw on a rich precedent, as poets, musicians and members of theatre groups like *Yuyachkani*. A preferred poetic form has been the *wayno,* whose roots go back to the Inca court. The political leader Lio Quintanilla chose the *wayno* to celebrate the taking back of stolen peasant lands in Andahuaylas in 1972; urging resistance in his hometown Huamanga (Ayacucho) in another piece in the same form, Eusebio Huamani decries the *sinchi* police, whose mottled green uniforms identify them as arrogant parrots that infest home and fields. A *wayno* of quite devastating power is '*Viva la patria*' by Carlos Falconi, which like Lienlaf's '*Rupamum*' uses the technique of incorporating Spanish words, this time into Quechua, in order to deconstruct and ultimately revile them, to the extent that the '*patria*' in question is exposed as vicious hypocrisy, an imposition both incoherent and insulting on all those who are not Latin or white:

Takichum takisqay wiqichum wiqillay	When the eyes of children

NATIVE AMERICA: POETRY

warmachakunapa ñawichallampi
Chiqnikuy huntaptin
Takichum takisqay wiqichum
y wiqilla

Vinchus viudalla asirillanmanchu

Cangallu viuda kusirikunmanchu

allqupa churinta unanchallanmanchu

pimanraq kutinqa sapan paloma
quru sacha hina mana piniyuq.

Sipillawaptimpas sayarimusaqmi
chakiytawiptimpas sayarimusaqmi

makichallaykita haywaykullaway

utqaymi purinay, qamllama allinlla
Huamanga del alma, hatarillasunmi!

- qawachan -

Vacaytaqa nakankutaq

radiuylaqa apankutaq
'cholo tu madre' niwankutaq
'viva la patria' niwankutaq
'viva la patria' niwankutaq.

fill with hate
can my song still be sung?
Can my lament still be
a lament?

The widow from
Vinchos, will she laugh?
The widow from Cangallo,
will she be happy?
The son of a dog, will she
love as her own?
Who will the lone dove rely on
like a sick tree, with no-one

Even if they kill me I will stand
even if they break my feet I
will stand
Reach me your hand of
solidarity I need to travel fast
See you soon, be well
Huamanga del alma, we will arise

- coda -

Their style is to slaughter
my cow
steal my radio
say 'your wog mother'
say '*viva la patria*'
say '*viva la patria*'

The question of racial conflict and of identity within the nation-state recurs in Mexico where modern authors still, or again, turn to Nahuatl, the language once spoken at the courts of Tenochtitlana and Texcoco. This recuperation may involve no more than re-stating the aesthetics and philosophy of 'flower-song' *(xochi-cuicatl)*, that is, Nahuatl poetry itself; hence, Natalio Hernández Hernández's poem '*Nocolhua cuicate*' (Our ancestral singers) delicately revives the binary phrasing of the sixteenth century *Cantares mexicanos* manuscript in invoking the old capacity to 'say and know', 'say and sing'. Or, as in a poem by Fausto Hernández

Hernández, a traditional mode like the 'orphan song' (*icno-cuicatl*) may be employed to express the current predicament of children and families in Nahuatl-speaking Veracruz who have been abandoned by parents obliged to migrate to alien cities: the title '*Tototl*' ([migrant] bird) can refer to either gender, women having in fact borne much of this burden, earning money as they can in the hope of eventually helping those they left behind. In his '*Keski nauamaseualme tiistoke?*' (How many Nahua are we?), Luis Reyes laconically refers to official census figures for Nahuatl-speakers and the not-so-secret desire of the *coyotes* (whites) to see them dwindle:

Kenke, tle ipampa,	Why, for what reason
kitemojtokej matipoliuikan?	do they want us to disappear?
Ax moneki miak tiknemilisej	Not much thought is needed
se tsontli xiuitl techmachte	four hundred years have taught us
tlen kineki koyotl.	what coyote wants.
Koytl kieleuia totlal	*Coyote* fancies our land
kieliuia tokuatitla	fancies our woods
kieleuia toateno	fancies our rivers
kieleuia tosiouilis	fancies our labour
kieleuia toitonalis.	fancies our sweat.
Koyotl kineki matinemikan	*Coyote* wants us to live
uejueyi altepetl itempan	in the slums of big cities
nupeka matixijxipetsncmikan	there to live naked
nupeka matiapismikikan	there to starve to death
nupeka matokamokajkayauakan	there to become objects of their deceit
nupeka matokamauiltikan.	there to become objects of their game.
koyotl kineki matimochiuakan	*Coyote* wants us on his payroll
tiitlakeualuan.	
Yeka kineki matikauakan	That's why he wants us to give up
tokomontlal	our communal lands
tokomonteki	our communal labour
tomaseualteki	our native tasks
tomaseualtlajtol	our native speech
yeka kineki matikilkauakan	that's why he wants us to forget
tomaseualtlaken	our native clothes
tomaseualnemilis	our native way of life

NATIVE AMERICA: POETRY

tomaseuallalnamikilis.	our native way of thought.
Koyotl achto techkoyokuepa	First *coyote* turns us into *coyotes*
uan teipa techtlachtekilia	then he steals from us
nochi tlen touaxka	all that is ours
nochi tlen titlaeliltia	all that we produce
nochi tlen mila tlaelli	all that the *milpa* produces
kichteki tosiouilis	he steals our fatigue
kichteki totekipanolis.	he steals our work.

WORKS BY Dzul Poot, Paulino Yama and other modern Maya authors in Yucatan carry forward a literary tradition which, in celebrating Maya polity, stretches back unbroken over 1,500 years or more to the hieroglyphic texts of the classic-period cities. At the hard interface in Chiapas, Petu' Krus writes in Tzotzil Maya about survival as a woman, while in neighbouring Guatemala, Rigoberta Menchú shows in her autobiography how culture continues to be sustained by the cosmogony and beliefs recorded in the *Popol vuh*, the sixteenth-century classic in her language (Quiché-Maya). At the same time, besides Nahuatl and Maya, other Mesoamerican languages like Zapotec and Otomi (Ñahnu), which belong to the ancient Otomanguan family, are becoming better known through alphabetic texts. One such is Thaayrohyadi Bermudez's '*Tsi Mahkitee Lerma*', a heartfelt Otomi ode to the 'father-river' Lerma which denounces pollution in political and cosmic terms and passes on its ecological message by honouring the old water gods.

A decisive factor in these examples — Mapuche, Quechua, Nahuatl, Maya, Otomi — is the principle of continuity, of knowable history whose beginnings long antedate Columbus and which, in the last instance, is inseparable from the vaster story of the world ages or 'suns' told in the *Popol vuh*. Native American tenacity and resilience in the face of such massive assault, and apparently against all technological odds, argue for belief that is both practicable and renewable. As the Quito Declaration tells us, its source lies in cosmogony, in ancient yet modern accounts of how the earth was and still is being formed and how we as a species have come to inhabit it. ❏

Gordon Brotherston *is professor of Spanish and Portuguese at the University of Indiana at Bloomington and research professor at the department of literature at Essex University, UK*

HUMBERTO AK'ABAL

Poems

The following poems, written in Mayan, were collected by Siobhan Dowd of Pen American Center during a trip to Guatemala in 1994. She says, 'All were written against great odds, in situations of fear and repression. All seek to give voice to Mayan identity, itself repressed for so many centuries'

Paradise

PARADISE was here.
Corn, wheat, beans.
No forbidden fruit
and no talking serpents.

Jalic-chumil and Cowilajchee
made love in the meadows
and covered themselves with sky.

Then one day the serpents spake:
'Forbidden to eat fruit,' they said
and divided Paradise among
    themselves.

Tale

AN ancient people
tired of their language
— or so it was told —
decided to raise a hill.
They piled up earth
as high as the clouds.

Up there — it was told —
they handed out languages.

I wanted like mad to go up there —
you had to knock back a few strong
    drinks,
and you came down talking nonsense
in another language!

Humberto Ak'abal was born in Momostenango, Totoniapán, Guatemala, in 1952. His published works include El Animalero *(Cultura, 1990) and* Poems *(Artesanal, 1992), and have also appeared in journals and magazines in Guatemala, Mexico, the USA and France*

Translated by Sarah Arvio for PEN American Center

GUILLERMO MARTÍNEZ

Guillermo Martínez was born in Bahía Blanca, in the province of Buenos Aires, Argentina, in 1962, four years before General Onganía came into power. In 1982 he was awarded the first prize in the National Short Story Competition 'Roberto Arlt' for his book La jungla sin bestias *(The Beastless Jungle)*; six years later he received the first prize from the Fondo Nacional de las Artes for his second collection of short stories, Infierno Grande *(Vast Hell)*. His first novel, Acerca de Rodorer *(Concerning Rodorer)* was published in 1992.

Martínez belongs to the generation of writers who grew up in the midst of the Argentina of the 'dirty war' between the military dictatorship and the guerrilla, a war that left the country shattered and from which Argentina has not recovered in spite of the present government's attempts to erase all memory of those past atrocities. The war did violence to everyone and everything, including the Argentine language. The writers of Martínez's generation were forced to reconstruct a tongue destroyed by the abuse of power, by irrational violence, by forced stupidity which infected words like a virus infects the blood. Their task was not only to bear witness and to build imaginary landscapes for their chronicles which are not, it must be said, mere documentaries. First they had to rescue the words themselves from debasement, using a pared-down, clear-cut language, free from the rhetoric, far-fetched metaphor and bombast so dear to the military heart.

Alberto Manguel

A small town is a vast hell — *Argentinian proverb*
For Ariel, Julio, Marcelo and all the gang

OFTEN, when the grocery store is empty and all you can hear are the buzzing flies, I think of that young man whose name we never knew and whom no-one in town ever mentioned again.

For some reason which I can't explain, I always imagine him as we saw him that first time: the dusty clothes, the bristling beard and especially the long dishevelled hair that almost covered his eyes. It was the beginning of spring, which is why when he came into the store I took him for a camper headed south. He bought a few tins and some *maté* or coffee; as I added up his bill, he looked at his reflection in the window, brushed his hair off his forehead and asked me if there was a barber in town.

In those days, there were two barbers in Puente Viejo. Now I realise that if he'd gone to Old Melchor's he might never have met the French Woman and no-one would have gossiped. But Melchor's place was at the other end of town and, in any case, I don't think that what happened could have been avoided.

Cervinio was giving him a haircut, the French Woman appeared. And the French Woman looked at the boy the way she looked at all the men. And that was when the bloody business started, because the boy stayed on in town and we all thought the same thing: that he'd stayed on because of her.

It hadn't been a year since Cervinio and his wife had settled in Puente Viejo and what we knew about them was very little. They weren't sociable with anyone, as the whole town used to point out angrily. The truth be told, in poor Cervinio's case it was little more than shyness, but maybe the French Woman was, in fact, quite stuck-up. They'd come from the big city; they had arrived last summer, at the beginning of the season and, when Cervinio opened his barber shop, I remember thinking that he'd soon bring Old Melchor under, because Cervinio had a hairdresser's diploma and had won a prize in a crew-cutting competition, and owned a pair of electric clippers, a hair-dryer and a swivelling chair, and he would sprinkle vegetable extracts on to your scalp and would even spray some lotion on you if you didn't stop him on time. Also, in Cervinio's shop there was always the latest sports magazine in the rack. And, above all, there was the French Woman.

I never quite knew why they called her the French Woman, and I never tried to find out: I'd have been disappointed to discover that, for instance, the French Woman had been born in Bahía Blanca or, even worse, in a little town like this one. Whatever the truth, the fact is that I'd never before met a woman quite like her. Maybe it was simply that she didn't wear a bra: even in winter one could see that she wasn't wearing a thing under her sweater. Maybe it was her habit of appearing barely dressed in the barber shop and putting on her make-up in the mirror, right there in front of everyone. But that wasn't it: there was in the French Woman something even more disturbing than her body which always seemed uneasy in its clothes, even more unsettling than the low plunge of her neckline. She would stare you in the eyes, steadily, until you had to look down, and her eyes were full of incitement, full of promise, but also with a mocking glimmer, as if the French Woman were testing us, knowing in advance that no-one would take up her challenge, as if she had already made up her mind that no-one in town measured up to her wild standards. So she'd provoke us with her eyes and scornfully, also with her eyes, she'd draw away. All this in front of Cervinio who seemed to notice nothing, bent in silence over the back of our necks, clicking from time to time his scissors in the air.

Oh yes, the French Woman was at first Cervinio's best publicity and, in the early months, his barber shop was very busy. But I had been mistaken about Melchor. The old man was no fool and he gradually started to lure his clients back: he somehow managed to get some porno magazines which the military in those days had forbidden, and later, during the World Cup, he gathered all his savings and bought a colour TV, the first one to appear in town. Then he started saying, to whoever would listen, that in Puente Viejo there was one and only one barber shop for men; Cervinio's was a hairdresser's for poofs.

However, I believe that if many returned to Melchor's barber shop, it was, once again, because of the French Woman; there aren't many men who will bear for very long a woman who humiliates or makes fun of them.

As I was saying, the young man stayed on. He set up his tent in the outskirts of town, behind the dunes, not far from the house of Espinosa's widow. He rarely came to the grocery store; whenever he did, he'd buy groceries for a long haul, for a fortnight or a month, but every single day he'd visit the barber's. And, since it was hard to believe that he went there with no other purpose than to read the sports pages, people started feeling pity for Cervinio. Because that is what happened: in the beginning, everyone felt sorry for Cervinio. The truth is that it wasn't difficult to feel sorry for him: he had the innocent air of a cherub and an easy smile, as shy people often seem to have. He was a man of very few words and at times he appeared to sink into a tortuous and distant world; his eyes would wander into space and he'd stand for a long while sharpening his razor blade or interminably clicking his scissors, so that you had to cough to bring him back to reality. Once or twice I had surprised him in the mirror, staring at the French Woman with dumb, concentrated passion, as if he himself wasn't able to believe that such a woman was his wife. And we were filled with pity by his devoted gaze that held not the shadow of a doubt.

On the other hand, it was equally easy to condemn the French Woman, above all for the town's married women and for the spinsters in search of a husband who, from the very start, had made common cause against her fearful necklines. But also many men felt resentful against the French Woman: in the first place, those who had a reputation as the lady-killers of Puente Viejo, such as Nielsen the Jew: men who weren't accustomed to being slighted and, even less, to being scorned by a woman.

And either because the World Cup was over and there was nothing left to talk about, or because there was a dearth of scandals in town, all conversations ended up discussing the goings-on of the French Woman and her young man. From behind the counter, I'd hear over and over the same comments: what Nielsen had seen one night on the beach — it had been a cold night and yet they both stripped naked and they must have been on drugs because they had done something which Nielsen would not describe, even alone among the men; what Espinosa's widow had said — that from her window she could always hear laughter and moaning coming from the boy's tent, the unmistakable sound of two bodies rolling around together; what the eldest of the Vidal had told us — that right in the barber shop, right there in front of him and of Cervinio... Who knows how much of all that gossip was true.

ONE DAY we realised that the boy and the French Woman had both disappeared. I mean, the boy didn't seem to be around anymore, and no-one had seen the French Woman either in the barber shop or on the pathway down by the beach where she liked to go for walks. The first thing we all thought was that they'd run away together and, maybe because running away always has a romantic ring, or maybe because the dangerous temptress was now out of reach, the women seemed willing to forgive the French Woman. It was obvious that there was something wrong in that marriage, they'd say; Cervinio was too old for her and also, the boy was very handsome... And with secretive giggles they'd confess that maybe they would have done the same.

One afternoon, when the matter was being discussed once again and Espinosa's widow was in the grocery store, the widow said in a mysterious voice that in her opinion something far worse had taken place; the boy, as we all knew, had set up his tent near her house and, even though she, like all of us, hadn't seen him since, the tent was still there and it seemed to her very strange — she repeated the words, *very strange* — that they would not have taken the tent with them. Someone said that maybe the police should be told and then the widow muttered that it might also be convenient to keep an eye on Cervinio. I remember becoming angry and yet not knowing what to answer: my rule is never to enter into an argument with a client. I began by weakly saying that no-one should be accused without proofs, that in my opinion it was impossible that Cervinio, that someone like Cervinio... But the widow cut in: it was a well-known fact that shy people, introverted people, can be extremely dangerous when pushed too far.

We were still going round in circles when Cervinio appeared at the door. There was a deep silence; he must have realised that we were talking about him because everyone tried looking in other directions. I saw him blush and, more than ever, he seemed to me like a helpless child who had never attempted to grow up. When he gave me his order I noticed that he had asked for only a few groceries and that he hadn't bought any yoghurt. While he was paying, the widow abruptly asked him about the French Woman. Cervinio blushed once more, but gently now, as if feeling honoured by so much solicitude. He said that his wife had travelled up to the city to look after her father who was very sick, but that she would soon be back, maybe in a week's time. When he finished speaking, a curious expression, which at first I found hard to define, had crept over all the faces: disappointment. And as soon as Cervinio was gone, the widow renewed her attack. She, said the widow, had not been taken in by that humbug: we'd never see the poor woman again. And in a low voice she insisted that there was a murderer on the loose in Puente Viejo, and that any one of us might be the next victim.

A WEEK went by, a whole month went by and the French Woman hadn't returned. Nor had the boy been seen again. The kids from town started using the tent to play at cowboys and indians, and Puente Viejo divided itself into two camps: those who were convinced that Cervinio was a criminal and those of us who believed that the French Woman would come back — and we were becoming fewer and fewer. One could hear people say that Cervinio had slit the boy's throat with a razor while cutting his hair, and mothers would forbid their children to play on the street outside the barber shop and would beg their husbands to go back to Melchor's. However, and this may seem strange, Cervinio wasn't left bereft of clients: the boys in town would dare one another to go sit in the doomed barber's chair and ask for a razor haircut, and it became a sign of virility to wear one's hair brushed upwards and sprayed.

When we'd ask for news of the French Woman, Cervinio would

repeat the story about the sick father-in-law, which no longer sounded believable. People stopped greeting him and we heard that Espinosa's widow had told the police inspector that he should be arrested. But the inspector had answered that until the bodies were found, nothing could be done.

The town started making conjectures about the bodies: some said that Cervinio had buried them in his patio; others, that he'd cut them into strips and thrown them into the sea. And gradually, in the town's imagination, Cervinio grew into an increasingly monstrous being.

In the grocery store, listening constantly to the same talk over and over again, I began to feel a superstitious fear, the presentiment that in these endless discussions something awful was being hatched. In the meantime, Espinosa's widow seemed to have gone out of her mind. She went about digging holes everywhere, armed with a ridiculous children's spade, hollering at the top of her voice that she wouldn't rest until she'd found the bodies.

And one day she found them.

IT WAS AN AFTERNOON at the beginning of November. The widow came into the store and asked me if I had any shovels and then, in a loud voice so that everyone would hear, she said that the inspector had sent her in search of shovels and of volunteers to dig in the dunes behind the bridge. Next, slowly dropping the words one by one, she said that it was there that she had seen, with her very own eyes, a dog devouring a human hand. A shiver ran down my back; suddenly, it had all become true and while I was looking for the shovels and while I locked up the store, I kept on hearing, without quite believing it yet, the horrible conversation: 'dog', 'hand', '*human* hand'.

Proudly, the widow led the march. I trailed behind, in the rear, carrying the shovels. I looked at the others and saw the usual faces, the people who came to the store to buy pasta and tea. I looked around me and nothing had changed, no sudden gust of wind, no unexpected silence. It was an afternoon like all others, at that useless hour at which one wakes up from one's nap. Below us, the houses stood in an ever-decreasing line, and the sea itself, in the distance, seemed provincial, unthreatening. For an instant I thought I understood my own feelings of incredulity. Because something like this couldn't be happening here, not in Puente Viejo.

WHEN we reached the dunes, the inspector had not yet found anything. He was digging bare-chested, and his shovel rose and fell unhampered. Vaguely he pointed around and I handed out the shovels, and sunk mine in the spot that looked safest. For a while, all that was heard were the dry thuds of the metal hitting the sand. I was losing my fear of the shovel and was thinking that maybe the widow had made a mistake, that maybe what she had told us wasn't true, when we heard a furious barking.

It was the dog the widow had seen earlier, a poor anaemic creature running desperately in circles around us. The inspector tried to shoo it away by throwing bricks at it, but the dog came back again and again, and at a certain point seemed almost to jump up at the inspector's throat. And then we realised that this was indeed the place. The inspector started to dig once again, faster and faster; his frenzy was contagious, the shovels dug in all together and suddenly the inspector shouted that he'd hit something. He dug a little deeper and the first body appeared.

The others barely threw a glance at it and went back to their shovels, almost enthusiastically, searching for the French Woman, but I went up to the body and forced myself to look at it closely. It had a black hole in the forehead and sand in the eyes. It wasn't the boy.

I turned around, to warn the inspector, and it was like stepping into a nightmare: they were all digging up bodies. It was as if the bodies were sprouting from the earth: every time a shovel dug in, a head would roll out or a mutilated torso would appear. Wherever you looked there were dead bodies and more dead bodies, and heads, and more heads.

The horror made me wander from one place to another; I wasn't able to think, I wasn't able to understand, until I saw a back riddled with bullets and further away a blindfolded head. Then I realised what it was. I looked at the inspector and he too had understood, and he ordered us to stay where we were, not to move, and went back into town to ask for instructions.

Of the time that went by until he came back, I only remember the incessant barking of the dog, the smell of death and the figure of the widow prodding with her children's spade among the corpses, shouting at us to carry on, that the French Woman had not yet been found. When the inspector returned, he was walking straight-backed and solemn, like someone ready to give orders. He stood in front of us and told us to bury the bodies again, just as we had found them. We all went back to our shovels, no-one daring to say a word. While the sand covered the bodies, I asked myself whether the boy might not be here as well. The dog was barking and jumping up and down, as if crazy. Then we saw the inspector, one knee on the ground and his gun in his hand. He fired a single shot. The dog fell down dead. Then he took two steps still holding the gun and kicked the dog's body away, for us to bury it as well.

Before going back, he ordered us not to speak to anyone about this, and jotted down, one by one, the names of all of us who'd been there.

THE French Woman returned a few days later: her father had completely recovered. We never mentioned the boy again. The tent was stolen as soon as the holiday season started. ❏

© *Translated by Alberto Manguel; illustrated by Oscar Grillo*

LEGAL

A legal column dedicated to the memory of Bernie Simons (1941-1993), radical lawyer and defender of human rights

JULIAN PETLEY

Savoy scrapbook

In a liberal society, issues of civil liberties and free expression are generally fought out on the margins of public debate, as Savoy publishers discovered when their *Lord Horror* ran foul of the law

'THE WORST CASE on record [of harassment by local vice squads] is the 20-year persecution of David Britton and Michael Butterworth, the Manchester-based publishers, whose authors have included Michael Moorcock and Jack Trevor Story, by the Manchester Obscene Publications Squad. David Britton first went to prison in 1982 for possessing "obscenity" — Grove Press and Venus Freeway novels, which were openly on sale in every other part of the country. Eleven years later, he was jailed again for publishing his own novel, Lord Horror. The real reason for this scandal was their company's open opposition to the arbitrary powers of Chief Constable James Anderton, whose satirical characterisation in Lord Horror led to the second conviction. Britton and Butterworth funded their publishing enterprise through several local bookstores selling paperbacks, American comics, minority interest books and periodicals, and soft-core pornography. They have lost count of the number of raids that have taken place and the destruction orders under Section Three of the 1959 [Obscene Publications] Act, which even consigned the likes of *Penthouse* and *Mayfair* to the flames.'
Bill Thompson *Softcore: Moral Crusades Against Pornography in Britain and America*, Cassell 1994.

November 1980 The Savoy offices and all Manchester Savoy retail outlets are raided in a co-ordinated swoop by the Manchester police under their highly controversial chief constable, James Anderton. Amongst the thousands of pounds' worth of stock seized are Samuel Delaney's *Tides of Lust* and Charles Platt's *The Gas* (both cultish, erotic science fiction novels) and Jack Trevor Story's *The Screwrape Lettuce*.

February 1981 Savoy Books are forced into temporary liquidation by a combination of recession, the collapse of the New English Library (with whom Savoy had a useful tie-in) and five years of escalating police harassment.

'Since 1976, when James Anderton took over as chief constable, Savoy premises had been raided over 40 times, resulting in the loss of tens of thousands of pounds worth of stock. Between 1977 and 1981 Anderton obtained a total of 1,010 search warrants from magistrates under the Obscene Publications Act, which meant that on average one Manchester premises was being raided every two days. The material seized and destroyed was mostly soft-core pornography of the kind that could normally be distributed and sold elsewhere in Britain with impunity.'

Michael Butterworth, 'Under Siege', *Savoy Dreams: The Secret Life of Savoy Books*, David Britton and Michael Butterworth (eds), Savoy 1984.

May 1982 David Britton serves 19 days of the 28-day sentence imposed as a result of the police raid of November 1980.

1983 Michael Moorcock publishes the anti-censorship pamphlet *Retreat From Liberty* (Zomba) in response to the police campaign against Savoy.

May 1989 Savoy publish David Britton's novel *Lord Horror*. Its malevolent central character is a mythic recreation of 'Lord Haw-Haw', alias William Joyce, the wartime traitor hanged in 1946 for his infamous 'Germany calling' radio broadcasts. The novel concerns his post-war search for Hitler in a surreal, ultra-violent, post-modern world.

'*Lord Horror is a sci-fi novel of an "alternative universe" where all the events, characters and scenes are metaphorically playing out philosophical and metaphysical abstractions in a sequence of symbolic forms. This is David Britton's* Pilgrim's Progress — *or at least his* Childermass.

'*In a manner similar to the discourses of De Sade's* Philosophy in the Boudoir, *the dialogue consists of contrived argument and counter-argument, encapsulations of every major train of thought and belief that has made the twentieth century the horror we see today — the characters voicing all the insane dialectic which has fuelled the nightmare of western culture. Lord Horror himself is an extreme aesthete — a psychopathic/neuropathic dreamer — a cross between Des Esseintes and Darth Vader. He is here likened, in this respect, to Hitler, who also (the book suggests) dreamed of higher things, of beauty, purity and glory divorced from reality. Horror is the epitome of Hitler's version of the Ubermensch — amoral, physically powerful and ruthless, agonisingly hypersensitive and mystically inclined, with a violent scorpionic sexuality. He is a Byronic anti-hero, his goals super-human, his actions sub-human.*

'*The exaggerations, the surreal*

imagery, and the distorted misappropriation of historical characters actually define a vision closer to the truth than mere "social realism" would ever be able to, revealing the corrupted inner life of characters, things and events — the dreaming reality of the historical process.'

'Black Easter: The Trials of Savoy Books' by Paul Anthony Woods and D M Mitchell, in *Rapid Eye 2*, Simon Dwyer (ed), Creation Books 1995.

June 1989 The start of Savoy Comics, with *Lord Horror* No 1 and *Meng and Ecker* No 1 (short for Josef Mengele and Dietrich Eckart, as well as a reference to the name of a Manchester cafe).

September 1989 In a front-page article in the *Jewish Telegraph*'s Manchester edition, Doreen Wachmann points out that, in the novel *Lord Horror*, James Anderton's notorious speech condemning homosexuals has been put into the mouth of a character named James Appleton — the only difference is that the word 'homosexuals' has been replaced by the word 'Jews' throughout. Wachmann raises the matter with Anderton, who is quoted as saying that he is 'looking into' it. The *Manchester Evening News* picks up the story. Shortly thereafter police raid Savoy's offices and shops, acting under Section Three of the Obscene Publications Act and carrying warrants signed by stipendiary magistrate Derick Fairclough. They seize all material related to *Lord Horror*.

May 1990 In an article by Paul Ferris in the *Observer,* two anonymous Manchester police officers complain about not being able to bring charges under Section Two of the Obscene Publications Act against the authors and publishers of a novel which, although not named, is clearly *Lord Horror*. Section Two is more serious than SectionTthree in that it carries a criminal penalty; however, Section Three is routinely used to put publishers out of business since it empowers magistrates to destroy their stock without the allegedly offending material even going before a jury.

August 1991 The Section Three charge is heard by Fairclough who finds the novel *Lord Horror* and *Meng and Ecker* No 1 obscene and orders their destruction. Savoy appeals. Three days later, and again using warrants signed by Fairclough, the police raid Savoy and seize over 4,000 *Lord Horror*-related items.

July 1992 Article 19 having taken up Savoy's case and drawn it to the attention of Geoffrey Robertson QC, he defends them at their appeal at Manchester Crown Court against the August 1991 conviction. Expert witnesses called in Savoy's defence include Michael Moorcock, Guy Cumberbatch and Brian Stableford. Moorcock argues that *Lord Horror* 'is in a tradition of lampoon, of exaggeration. Its purpose is to show up social evils, and the evils within ourselves. The book tries to identify the ways of thinking that led to the Holocaust, and could yet lead to another one', whilst Stableford places it squarely within 'a tradition running from the Decadent and Symbolist fiction of the 1890s, through to the Surrealist movement'. Robertson makes a telling contemporary point by noting that 'unlike the *Sunday Times* publication of Goebbels' diaries, this work has no appeal to neo-Nazis... No Nazi skinhead-type would get past page two. The Holocaust cannot be excluded from the literary imagination on his account.'

Judge Gerard Humphries lifts the destruction order against *Lord Horror*, noting that 'no-one is prepared to read this work unless they are willing to digest large amounts of philosophy and complex argument,' although adding, 'we give this book no accolade, no approval.' The charges against *Meng and Ecker* No 1, however, are upheld, the judge declaring it

Front cover of **Meng & Ecker**, *number 5, January 1992*

to be 'luridly bound' and thus far more likely than the novel 'to attract attention from the less literate'. He also states that it is 'a glorification of racism and violence. It contains pictures that will be repulsive to right-thinking people, and could be read — and possibly gloated over — by people who enjoy viciousness and violence. We do not consider it in the public interest that it should be put on sale.' Counsel advise Savoy that there are no reliable precedents for taking a section three seizure case to a higher court.

April 1993 David Britton is sentenced to four months' imprisonment for the sale of non-Savoy material seized in the second of the August 1991 police raids.

June 1995 Savoy, their case now handled not only by Robertson's chambers but also Stephens Innocent, go to Manchester Magistrates Court to test the 1964 undertaking given by the solicitor-general to Parliament that publishers of serious works caught in Section Three seizure trawls for pornography will (unlike Savoy) be allowed to opt for trial by jury if they so desire. The case arises from the August 1991 seizure of *Lord Horror* material. The magistrate is switched at the last moment from Derick Fairclough to Jane Hayward. The latter agrees that Savoy have a 'legitimate expectation' of a jury trial, but sides with the prosecution's claim that the local Crown Prosecution Service holds a discretion. The local CPS consider whether Savoy are serious publishers, but decide they are not. Hayward refuses to press for substantiation of what is only an 'opinion' and insists on hearing the case despite defence counsel's request that she abandon it, and the defence protestation that they intend to take the matter to judicial review. The prosecution invokes the increasingly discredited and dubious Public Interest Immunity legislation to hide its reasons for disallowing a jury trial, thus making impossible any discussion or analysis of its grounds for stating that Savoy are not serious publishers.

The case is adjourned until the following month, when Hayward finds all the *Lord Horror*-related graphic titles obscene. She further finds that they do not attract a 'public good' defence under Section Four of the Obscene Publications Act but, indeed, are of a 'deliberately offensive, racist nature [without] literary, artistic or educational merit.'

October 1995 A copy of *Lord Horror* fetches £220 in *Index*'s Auction of Banned Books (p56). The following day Savoy appear before Judge Stephen Sedley at the Royal Courts of Justice and are granted leave for a judicial review of Hayward's refusal to state a case. Faced with this threat Hayward caves in and and agrees to state her case within 21 days. (Her draft is expected at the time of writing). ❑

Julian Petley is head of Communication and Information Studies at Brunel University, UK

INDEX INDEX

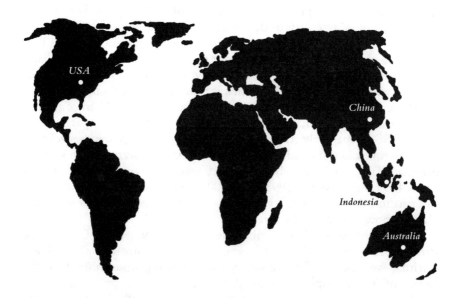

Breasting the tide of filth

IN DECEMBER last year, reports that the USA's largest Internet provider, America Online, had banned its subscribers from using the word 'breast' provoked understandable outrage from — among others — a group of women who used the service to talk to each other about their experience with breast cancer. If a subscriber tried to use the word, an automatic message would come back telling them their material was 'vulgar'. AOL quickly realised they had made a big mistake and reversed the policy. But the case pointed out the essential silliness of much of the debate about who should be policing the information superhighway and how.

In the USA, the 'online smut' debate now has less to do with genuine concern for children innocently logging on to bulletin boards run by paederasts or zoophiles (see p80). Instead it is displaying the tone of moral panic associated with a sure-fire vote-winner. Congressional discussions on how to deal with some of the Internet's more egregious outgrowths have been lumbering on month after month and, with elections approaching, it is increasingly difficult for any decent-minded representative of the people to be seen to be on the side of the pornographers. Hence the decision of the House-Senate conference committee in December — after months of lobbying by free speech advocates, and despite serious doubts about its legality — to return to an earlier, far more restrictive version of the Communications Decency Bill. The latest version, which is now likely to be passed into law, will hand down prison sentences and fines to anyone who transmits indecent material without first making sure that it is not accessible by minors.

The USA is not the only country exercised by the problem of taming this anarchic domain. In Australia, the Victoria state parliament introduced a remarkably stringent set of regulations in November relating to the transmission of child pornography and what it termed 'vile sadistic' material. And in China, where Internet access has grown exponentially (from 400 hosts in March 1995 to 12,000 by the end of the year) the State Information Centre has let it be known that as yet it has no workable ideas on how to control the Internet.

In a toughly worded editorial in December, however, the Communist Party paper *Renmin Ribao* said it will not countenance the physical and spiritual well-being of the country's youth being undermined by the spread of 'pornographic poison'. Undoubtedly there are a number of legislators in the USA who would happily concur with that formulation of the problem.

In Indonesia the problem is more explicitly political. Information minister Harmoko has declared, with commendable frankness, that his dislike of the Internet has less to do with pornography than the threat of spreading political dissent. It is ironic, then, that Goenawan Mohamad, who has now won two court cases against Harmoko in an attempt to get his magazine *Tempo* unbanned (*Index* 4&5/1994, 3/1995), has announced his intention of becoming the first dissident to seek asylum in cyberspace. He plans to launch a news service on the Internet for Asian journalists who want to avoid censorship at home.

Back in China, meanwhile, the Internet's political nuances have certainly not gone unnoticed. The Chinese government is reported to want to change Taiwan's Internet domain name from '.tw' to '.tw.cn', thereby achieving the virtual annexation of its recalcitrant former province. That has to be one of the subtlest forms ever devised of pursuing a territorial claim. ❑

Adam Newey

INDEX INDEX

A censorship chronicle incorporating information from the American Association for the Advancement of Science Human Rights Action Network (AAASHRAN), Amnesty International (AI), Article 19 (A19), the BBC Monitoring Service Summary of World Broadcasts (SWB), the Committee to Protect Journalists (CPJ), the Canadian Committee to Protect Journalists (CCPJ), the Inter-American Press Association (IAPA), the International Federation of Journalists (IFJ/FIP), Human Rights Watch (HRW), the Media Institute of Southern Africa (MISA), International PEN (PEN), Open Media Research Institute Daily Digest (OMRI), Reporters Sans Frontières (RSF) and other sources.

ALBANIA

On 27 September the former editor of *Lajmentari*, Gjergji Zefi, was convicted of defamation and forbidden from holding public office or writing publicly for one year, in connection with two articles he published in 1993 and 1994. Under the press law, editors are held accountable for publishing allegedly slanderous material, but the first article, which accused a government official of corruption, was published before the press law was enacted. (HRW)

On 28 October Felix Bilani, a photographer with the main opposition paper *Koha Jone*, was detained by Tirana police for taking photographs of a murder scene. His film was also destroyed. On 1 November dynamite exploded at the home of the paper's publisher Nikolle Lesi (*Index* 5&6/1993). No-one was hurt but the blast caused extensive damage. (HRW, SWB)

Parliament passed a law on 30 November allowing selective access to the files of the Sigurimi, the secret police under the Communist regime. Under the Law on the Verification of the Stature of Political and Public Figures, a committee will examine the files of all those seeking public office, as well as those working in the state media and at newspapers with a daily circulation above 3,000. Those found to have collaborated with the Sigurimi will be barred from office until 2002. The files of all other citizens will be closed for 30 years. (OMRI, A19)

On 5 December Blendi Fevziu, editor of the opposition paper *Aleanca*, was found guilty of defamation and fined US$2,000. The charges related to an article, published in April, accusing Blerim Cela, the State Control Service chief, of corruption, based on information already made public in Parliament and on state television. Fevziu had been planning to run as an opposition Democratic Alliance candidate in parliamentary elections due to take place later this year. (HRW)

ALGERIA

A car bomb outside the offices of *Le Matin* newspaper in Hussein Dey, Algiers, was defused by bomb disposal experts 20 seconds before it was due to explode on 30 October. Staff had already been evacuated from the premises by security forces. (Reuter)

A number of political groups, including the Islamic Salvation Front (FIS) and the Berber Cultural Movement (MCB), boycotted the presidential election on 16 November which was won by the incumbent, Liamine Zeroual. The Ministry of Communication warned the media against publishing calls for a boycott of the election. Parties opposed to the election were refused permission to hold meetings and a number of MCB members were arrested on 12 November. (HRW)

Three editions of French-language weekly *La Nation* were confiscated by security forces during the run-up to the election. The paper's managing editor and one journalist have been on probation for three months following the publication of an interview with FIS leader Abdelkader Omar. The editor of *El Watan*, Omar Belhouchet (*Index* 8&9/1993), is also on probation as a result of a story alleging corruption in the Health Ministry (*Index* 3/1994). (SWB)

The financial director of the Algerian Press Agency (APS), Ahmed Khalfoun, was shot dead near his home in Algiers on 6 November. Hamid Mahiout, a journalist with the French-language daily *Liberté*, and his driver Ahmed Belkherfallah, were kidnapped in Algiers on 2 December. Their decapitated bodies were found the next day. And on 5 December Khedidja Dahmani, a reporter for the independent Arabic-language weekly *Echourouk al-Arabi*, was shot dead near her home in the Algiers suburb of Baraki. (Reuter, SWB, CPJ)

ANGOLA

Mario Paiva, a Luanda-based freelancer for Reuters and the South African Broadcasting Corporation's Channel Africa, received death threats from a state security agent on 18 November. Paiva also claims his house is being watched by security agents from the Ministry of Home Affairs. (MISA)

AUSTRALIA

The influential Senate Select Committee on Community Standards recommended on 27 October that no violent or sexually explicit material be shown on television before 10.30pm. The Australian Broadcasting Authority would be able to impose on-the-spot fines for breaches of the revised code, with broadcasters losing their licences if they do not pay. (Melbourne *Age*)

The Victoria state parliament passed the omnibus Classification (Publications, Films and Computer Games) (Enforcement) Bill on 23 November. The law's provisions regulating pornography on the Internet have been sharply criticised for being over-broad. Legal experts fear that service providers could be held liable for material that they unwittingly carry on their services. (Melbourne *Age*)

The film *Dead Man* by director Jim Jarmusch was refused classification by the Office of Film and Literature Classification on 29 November, because of a scene depicting a woman being forced at gunpoint to fellate a man. The film's local distributor plans to appeal the decision to the Film and Literature Board of Review. (*Daily Variety*, Melbourne *Age*)

The New South Wales attorney-general has proposed an amendment to the Evidence Act, which would allow journalists to protect their sources. Under the proposal, a court could excuse a journalist from answering questions or producing documents if to do so would breach a confidential relationship between journalist and source. Since 1990 at least six journalists have been punished for not revealing sources, two with imprisonment. (*Sydney Morning Herald*)

AZERBAIJAN

President Aliyev pardoned jailed *Cheshma* journalists Ayez Ahmadov, Yadigar Memdli, Malik Bayramov and Asker Ahmed on 12 November (*Index* 6/1995). He said that the court verdict was 'just', but that the four are 'young and will draw conclusions from this event'. (SWB)

BAHRAIN

Munira Ahmed Fahro, associate professor of sociology at the University of Bahrain, was suspended from her teaching position on 2 October after she refused to withdraw support for the Bahraini Women's Petition, which she signed in April 1995. The Bahraini Women's Petition calls for the restoration of the Constitution and Parliament, respect by security forces for the rights of demonstrators, and participation by women in the political process. The government has sought to punish the signatories, including many who held teaching posts at the University of Bahrain and in public schools. Fearing the loss of their jobs, some women have repudiated their support of the petition. (AAASHRAN)

BANGLADESH

Reuters Television cameraman Rafiqur Rahman was injured by shrapnel, and 70 other people were hurt in fighting with police during a transport blockade in Dhaka on 6 November. The protest was intended to force the government to resign and call a general election. (Reuter)

BELARUS

On 11 September the minister of culture and the press suspended the licence of the Chata publishing house for one year, for publishing the book *The Pahonia in Your Heart and Mine*, which, he said, runs counter to the results of the referendum of 14 May 1995. One of the questions in the referendum was whether to replace the 'Pahonia' — a charging knight, which was adopted as the state coat of arms following independence — by the old Soviet symbol. In spite of a manipulated campaign, only 47 per cent of the electorate voted for the change. Nevertheless President Lukashenka announced that the Pahonia, and also the white-red-white flag of independence, would be replaced. According to the minister's order, by praising the Pahonia, the book poses a threat to civic order. Ironically, the order was issued on official notepaper which still had the Pahonia on it. (Belarusian PEN)

The opposition papers *Belorusskaya Delovaya Gazeta* and its cultural supplement *Imya* (*Index* 6/1995) are now being printed in Vilnius by the printing works which produces the Vilnius daily *Lietuvos Rytas*. The Belarusian post office and the state-owned Sayuzdruk distrubution agency, however, are refusing to handle independent newspapers. *Belorusskaya Delovaya Gazeta* and the independent *Belaruskaya Hazieta*, which has been printed in Vilnius since September, are now trying to set up their own distribution agency. (SWB)

President Lukashenka issued a decree on 15 November splitting the Ministry of Culture and the Press into two separate ministries. Religious affairs, which were the responsibility of the combined ministry, will now come under a special Committee for Religious Affairs attached to the Council of Ministers, following the old Soviet model. (Belarusian press)

BOLIVIA

Samuel Doria Medina, president of the board of the La Paz-based daily *Hoy*, was kidnapped on 1 November by a group of unknown assailants. No group has claimed responsibility and the reasons for the abduction are unknown. (IAPA)

BOSNIA-HERCEGOVINA

On 29 October the body of Goran Pejcinovic, a driver for Bosnian Serb TV, was formally identified (*Index* 6/1995). Pejcinovic's body was one of eight returned by Bosnian government authorities in an exchange on 25 October (not 21 October as previously reported). The body of Sasa Kolevski, a Bosnian Serb TV cameraman, was not among those released, contrary to earlier reports. He remains unaccounted for. (CPJ)

On 8 November *Christian Science Monitor* journalist David Rohde was released from a Bosnian Serb prison in Bijeljina where he had been detained since 29 October on charges of illegal border-crossing and falsifying documents. Rohde's reports had exposed massacres of Muslims by Serbs in Srebrenica. On 15 November he reported that he had discovered mass graves in Sahanici before he was captured, and accused Bosnian Serb forces of tampering with evidence. (OMRI, CPJ, SWB)

KULTURA

BRAZIL

Caio Ferraz, the administrator of Casa da Paz (Peace House) in the Rio de Janeiro district of Vigario Geral, has been subjected to death threats and harassment by members of the police. Casa da Paz is the site of the murder of eight people by members of the military police in August 1993. It has since been turned into a meeting house for the relatives of the dead. Fears for the safety of residents of Vigario Geral, particularly members of Casa da Paz, are heightened by reports that in recent months armed police have made several incursions into the district and carried out spot searches in the Casa da Paz. (AI)

Quintín Vargas, Delfín Olivera, Esteban Quispe and Teodocio Romero, leaders of the Coca Leaf Growers' Confederations and Trade Unions, were arrested by the Mobile Rural Patrol Unit (UMOPAR) on 15 and 17 November in Shinahota and Entre Rios, El Chapare region. The arrests took place during joint operations by UMOPAR and security forces, in which scores of peasants, trade union leaders and other civilians were arrested and tear gas bombs were dropped from helicopters. A six-month-old boy died of asphyxiation allegedly as a result of tear gas, and a 13-year-old girl was killed during reportedly indiscriminate gunfire. (AI)

BULGARIA

It was reported on 11 November that Ivan Granitski, Bulgarian National TV director-general (*Index* 6/1995), is under investigation for violations of the provisional statute on state-run media. The Union of Democratic Forces has repeatedly accused him of denying its representatives access to air-time. Granitski could be dismissed if legal proceedings begin and would face up to five years in jail if convicted. (OMRI)

On 14 November Sofia newspaper vendor Varban Yordanov reported that two men had wrecked his newsstand and tried to kill him. The incident is believed to be connected to a distribution war over control of lucrative sales spots. A private company, Primex 89, has recently built new booths in the capital and it is alleged that gangsters have been intimidating the owners of the old booths in prime locations in an attempt to prevent them operating. Primex 89 denies any connection with the incidents. (Reuter)

On 22 November 34 Bulgarian National Radio journalists from the station Horizont accused the government of censoring the state media by interfering in editorial decisions over news selection, rearranging newscasts and virtually preventing journalists producing their own commentaries. Journalists face heavy fines for not complying with government regulations. On 28 November Georgi Vasilksi and Peter Kolev of BNR's *Hristo Botev* programme issued a statement of support for their colleagues, and were sacked the following day. (OMRI)

BURUNDI

On 4 November UNESCO director-general Federico Mayor promised to ask the International Telecommunications Union to examine the possibility of taking legal action against the pirate radio station Voice of Democracy, which is accused of inciting racial hatred in Burundi. President Ntibantunganya denounced the radio station in an address to the UNESCO general conference on 30 October. (SWB)

CAMBODIA

On 30 October, addressing residents of Kraingyov who carried out an attack on the offices of the daily *Serei Pheap Thmey* (*Index* 6/1995), the second prime minister, Hun Sen, told them they had done nothing wrong. 'I am not encouraging people to attack any other people, but if they go, go ahead and I won't stop them,' he said. 'If they [the newspaper] can exercise their rights, why can't Kraingyov people?' (Human Rights Task Force)

The *Cambodia Daily* reported on 7 November that several Khmer-language newspapers critical of the government may have to suspend publication because they are unable to find a printer willing to accept their business. . The papers concerned are *Voice of Khmer Youth*, *Khmer Conscience*, *Khmer Ideal*, *Morning News* and *New Liberty*

News. The editors believe the decision by the Panh Chak Por printing house not to continue printing the titles is due to political pressure. (Human Rights Task Force)

On 11 November the information minister, Ieng Mouly, asked the Phnom Penh municipal court to close the unregistered paper *Forget News*, which had published three issues since late October. The newspaper aroused controversy over its style of reporting and its criticism of the foreign presence in Cambodia. The 1995 Press Law requires all newspapers to register with the government before publication. (Human Rights Task Force)

CAMEROON

On 27 October Ndzana Seme (*Index* 4/1995), managing editor of *Le Nouvel Indépendant*, lost his appeal against a one-year prison term and a fine of 100 million CFA francs (US$350,000) for insulting the head of state and incitement to rebellion. Owing to a series of articles he published on homosexuality and drugs in prisons, Seme fears for his safety if jailed and is now in hiding. (RSF)

On 16 November, new legal proposals governing the press were debated in Parliament. Article 17 of the measure would allow seizure and banning of newspapers if they attack 'the public order' or 'accepted standards of good behaviour', with those terms left open to interpretation. Article 7 makes the process for establishing a newspaper more difficult than at present. (RSF)

CANADA

The development of v-chip television blocking technology for the Canadian market has led to the downplaying of the Canadian Radio-Television and Telecommunications Commission's threat to block out violent US broadcasts, which had already resulted in the banning of programmes such as *Mighty Morphin' Power Rangers*. CRTC chairman Keith Spicer commented that v-chip technology appeared to make enforced censorship less likely, and said that if v-chip legislation were passed in the US 'this would be wonderful news for Canada as we could then try to work out a common cross-border classification system.' (Multichannel News)

Recent publication: *Blindspots in the News?* (Project Censored Canada 1995 Yearbook, School of Communication, Simon Fraser University, Burnaby, BC, V5A 1S6, Canada, 62pp)

CHINA

Ding Zilin (*Index* 5/1995) was released on 5 October after nearly two months in detention. Her husband, Jiang Peikun, is believed to have been released at the same time. (AI)

On 26 October Wang Zhihong, wife of Chen Ziming (*Index* 6/1995), was released from custody, after being held for five days for taking part in a peaceful protest. She is now under virtual house arrest at her husband's parents' house. Chen's parents were also held for several days for their part in the protest. (AI, AAASHRAN)

In late October it was reported that China hopes to change Taiwan's Internet domain name from '.tw' to '.tw.cn'. The abbreviations 'tw' and 'cn' denote Taiwan and China respectively. (*Computing*)

Chinese police seized about 2,200 copies of banned publications promoting Tibetan independence, plus pirated pornographic material, in a raid on wholesalers in the city of Xian on 1 November. Eleven stores were found to be selling the illegal works. (Reuter)

On 21 November dissident and Nobel Prize nominee Wei Jingsheng was formally arrested and charged with 'conducting activities in an attempt to overthrow the government'. Wei was detained on 1 April 1994 and held incommunicado at an undisclosed location (*Index* 1&2/1994, 3/1994). During a brief period of freedom (he was released from a 15-year prison sentence in September 1993), Wei continued to call for democracy, spoke to foreign journalists, and wrote for overseas publications. On 13 December he was found guilty and sentenced to 14 years in prison. His political rights are also suspended for three years. (Reuter, Human Rights in China, RSF, HRW, PEN, *South China Morning Post*)

Massive crackdowns on illegal

publications and pornography in Henan and Zhejiang provinces were reported by Central Television on 23 November. Over 250,000 copies of illegal books and 163,000 pirated audio-visual products were confiscated, and 135 people arrested. (SWB)

China carried out large-scale military exercises on Dongshan Island, facing Taiwan, on 25 November. The mock invasion, using live ammunition and featuring tanks, battleships, and aircraft, was shown on Chinese state television. Taking place a week before Taiwan's parliamentary elections, the exercise has been interpreted as a warning against a pro-independence vote. It is the second time this year that China has used military display to indicate displeasure with its neighbour: in July and August it conducted missile tests off Taiwan's north coast, following Taiwanese President Lee Teng-hui's visit to the USA. (*Guardian*, Reuter)

Recent publication: *The World Bank and Forced Labor in China* (Laogai Research Foundation, October 1995, 22pp)

COLOMBIA

Civic leader Luis Tarazona Ortiz was abducted on 23 October by armed men in Barrio de Popa, Ocaña, Norte de Santander department, and has not been seen since. Eight heavily armed men who said they were from the attorney-general's office forced their way into his home. They threatened the family before ransacking the house and taking Tarazona away. Witnesses believe the abduction is the work of paramilitary groups which are operating in the area with the support of security forces. (AI)

The government has announced stiff penalties for broadcast media that violate censorship measures imposed under a limited form of emergency rule introduced on 2 November. The measures were imposed after the assassination of Alvaro Gómez, a former presidential candidate and outspoken critic of Ernesto Samper, in the capital. The decree bans the broadcast media from transmitting any interviews with or statements issued by members of leftist guerrilla groups or 'criminal organisations linked to subversion or terrorism'. It also prohibits them from identifying or disclosing the location of witnesses to a broad range of criminal acts including terrorism, sedition, kidnapping, extortion, and drug trafficking. Violation of the measures will be punishable by fines up to US$230,000 and the withdrawal of their licences for up to six months. (Reuter)

CONGO

A new press law was passed by the National Assembly on 22 November. Out of the law's 122 articles, 46 are dedicated to specifying penalties for violations. The law provides for prison terms of up to five years for journalists found guilty of defamation, and extends the defamation provisions to cover printers and distributors. New powers are granted to the government to 'requisition' newspapers that 'seriously disturb public order'. The law also states that journalists working for state media must be 'loyal' to the government. (RSF, SWB)

COTE D'IVOIRE

One demonstrator was killed on 17 October in the western town of Duoko, when security forces removed a roadblock set up in a pre-election protest. Opposition parties had called for a boycott of the presidential elections held on 22 October, as well as of parliamentary and municipal polls in November and December. (*Daily Observer*)

CROATIA

On 26 October Goran Flauder of the Osijek weekly *Bumerang* was beaten by a man, apparently a plainclothes police officer, while photographing members of the ruling HDZ as they prepared for an election rally. Uniformed officers coming to Flauder's aid seem to have been ordered away by the plainclothes officer, who then smashed the camera and beat Flauder as he tried to escape. (AI)

On 5 December it was reported that a showing of the anti-war film *Vukovar* at a forum on cultural diversity held in a United Nations auditorium was cancelled following protests by Croatia's deputy ambassador to the UN. The film, by US director Daniel North, was accused of being 'an affront to the innocent victims of Belgrade's aggression'. (Reuter)

CYPRUS

Twelve Limassol policemen, some of them senior officers, may face dismissal following accusations in a report commissioned by the government, which accuses them of torturing suspects between 1990 and 1992. The report, issued in November, described 11 cases in which detainees were treated in a manner 'reminiscent of the Middle Ages'. (Reuter)

CZECH REPUBLIC

The government approved a new bill regulating the media on 18 October. According to the culture minister, Pavel Tigrid, two sections of the original draft have been dropped: one on the right and obligation of journalists to keep their sources of information secret, and one on the obligation of state bodies to provide information to the media. Also omitted are two paragraphs on media ownership. 'Concentration of ownership' will, however, be monitored by the Ministry of Economic Co-operation, for which a special law will probably be passed, said Tigrid. (SWB)

EGYPT

Three journalists have been charged under the press law passed in May 1995. Magdi Hussein (*Index* 1&2/1994, 1/1995), editor of the Islamist opposition paper *al-Sha'ab*, is on trial accused of libel by the son of interior minister Hasan al-Alfi after running a story about an unnamed son of an unnamed minister refusing to pay the bill for a meal in a five-star Cairo hotel. Gemal Badawi, editor of *al-Wafd*, and Mohammed Abdel-Alim, an *al-Wafd* journalist, were accused of libel by a member of Parliament after a story on corruption appeared in *al-Wafd* in August. (Reuter)

Salah Abdel-Maksoud, a journalist with *al-Sha'ab* and a member of the Journalists' Association Council, was arrested on 20 October at Cairo airport and charged with membership of the banned Muslim Brotherhood. He was acquitted and released on 27 November by a military court. (Reuter, SWB)

Recent publications: *These Trials Are Illegal — Second Report by the EOHR on the Trial of Civilians Before Military Judges* (Egyptian Organization for Human Rights, June 1995, 24pp); *Silenced Voices — Second Report on Freedom of Opinion and Expression in Egypt* (EOHR, September 1995, 66pp); *Al-Taghreeb (Alienation) — The Third EOHR Report on the Conditions of Prisons in Egypt* (EOHR, September 1995, 28pp); *Deaths in Custody* (AI, October 1995, 10pp)

ETHIOPIA

In a statement issued on 1 December the Ethiopian Free Journalists Association called for the release of 13 journalists and managers of publishing houses who have been arrested in a fresh government crackdown. This brings the number of jailed journalists, including those already sentenced to prison terms, to 24. Four private newspapers — two published in Amharic and two in English — stopped printing in protest at the arrests. (Reuter)

Leading civil rights activist Mesfin Wolde-Mariam was reported to be facing trial on 11 December on charges relating to a report by the Human Rights Council (which he chairs) into the killing of several unarmed student demonstrators by troops in 1993. The charges carry a prison term of up to three years. (*Guardian*)

GAMBIA

At least 34 people, alleged to be supporters of the banned People's Progressive Party (PPP), were arrested on or around 12 October following a gathering by critics of the government on 10 October. A further six men were arrested in the days that followed, and are being held incommunicado at the Bakau military barracks near Banjul. Those arrested are accused of planning a demonstration in support of a return to power by ex-President Jawara. Apart from a decree banning political activities, under the Public Order Act no processions may be held without a licence. (AI, *Daily Observer*)

Addressing a rally on 28 October, Captain Yayha Jammeh, head of state and chairman of the Armed Forces Provisional Ruling Council (AFPRC), declared that journalists, freedom fighters and human rights activists should be 'got rid of'. The Gambia Press Union (GPU) has expressed deep concern over what it called 'the increasing hindrances the authorities have continuously placed in the way of journalists'. The AFPRC later said that Jammeh's remarks had been directed at 'enemies of this country disguised as journalists' and appealed to journalists to approach their work with objectivity and fairness. (*Daily Observer*, RSF)

It was reported in late October that Sierra Leonean journalist Chernor Ojukwu Sesay, working for the Gambian *Daily Observer*, had been expelled from Gambia without being given a reason. Sesay fled Sierra Leone in April 1995 after being accused of having links to the rebel Revolutionary United Front. (*West Africa*)

Tanya Domi, field director of the US-based National Democratic Institute (NDI), which was brought in to oversee Gambia's return to democracy, was expelled from the country on 7 November. Domi, who was appointed head of the NDI's Banjul office early in October, is said to have been expelled because she was 'not co-operating with the government'. She had been critical of the government's attitude to the media. (*Daily Observer*)

GEORGIA

On 13 October the Central Electoral Commission ruled that journalists standing for Parliament may be authors of programmes and publications only if they do not contain election campaign propaganda. President Edvard Shevardnadze's regular Monday interviews likewise may not contain campaign propaganda. (SWB)

Radio Russia reported on 12 November that Elizbar Javelidze, one of the leaders of the 21st Century-United Georgia bloc (which comprises supporters of former President Zviad Gamsakhurdia), had been arrested for making the 'anti-constitutional' statement that the organisation would do everything in its power to overthrow the present government. According to Interfax agency, the prosecutor's office has neither confirmed nor denied the reports of Javelidze's detention. (SWB)

GREECE

On 28 October Makis Psomiades, publisher of the daily *Onoma*, was sentenced to 16 months in jail on charges of 'unprovoked insult' for printing a front-page photograph allegedly showing Dimitra Liani, wife of prime minister Andreas Papandreou, in a naked pose with another woman. Under Greek law he can pay a fine instead of serving his sentence. Three other publishers and editors, including George Kouris, publisher of *Avriani*, face similar charges for publishing the same photographs. (Reuter)

On 14 November around 100 anarchists hurled petrol bombs at journalists covering a student march in Athens. A van from the private television station Skai was set on fire and a photographer was injured before police could intervene. (Reuter)

GUATEMALA

Human rights ombudsman Jorge García Laguardia's report on human rights for the first seven months of 1995 cites 10,233 denunciations of human rights abuses, including 159 extrajudicial executions, 57 disappearances, four cases of torture by state security agencies, and 141 abuses of authority. (*Mesoamerica*)

In July a team of international forensic experts completed exhumations of three sites in Los Dos Erres, Petén Department, where they recovered the remains of at least 162 people. During three days in December 1982, the army massacred an estimated 350 people in the area. As yet there has been no criminal investigation into this and other massacres that occurred during the army's scorched-earth policy in the 1980s. (AI)

On 16 and 17 October the Quiché home of Amílcar Méndez was surrounded by 30 to 40 soldiers. Méndez, president of the Council for Ethnic Communities (CERJ), and his family have been threatened and harassed on numerous occasions in the past. (AI)

Ernesto Bol, an executive member of the Municipal Workers' Union of Cobán, was beaten by six masked men on 20 November. The Labour Tribunal had recently ruled that Bol should be reinstated to the job from which he had been dismissed 20 months before. (AI)

On 3 December the body of a Mexican UN worker, who was kidnapped on 28 November, was found in a shallow grave in Quetzaltenango. Her body had three gunshot wounds and showed signs of torture. (AI)

Recent publications: *Victims of 1982 Army Massacre at Los Dos Erres Exhumed* (AI, October 1995, 16pp); *Maquila Workers Among Trade Unionists Targeted* (AI, November 1995, 14pp)

HAITI

On 7 November parliamentary deputy Jean Hubert Feuille, a cousin of President Aristide, was shot dead. Deputy Gabriel Fortune and a passer-by were seriously wounded in the same attack. Following Feuille's murder, protests erupted throughout Haiti. Dozens of homes were burned, and two former members of the Haitian army were stoned to death. Guns seized by members of the public were handed in to UN soldiers. Individuals and organisations have denounced the atmosphere of impunity, as well as the lack of disarmament, since the 1994 UN/US occupation. (*Latinamerica Press*, *Haiti Info*)

On 14 November protesters prevented the reopening of Radio Cap-Haïtien, which had ceased broadcasting earlier in the day after threats were made against their journalists. A senior editor, Jean Robert Lalanne, and a reporter, Carry Celestin, went into hiding after they were threatened with 'necklacing'. The threats are believed to be in connection with questions Lalanne asked the prime minister, Claudette Werleigh, regarding the participation of government supporters in violent disturbances in Cap-Haïtien. On 15 November the station's head, Evelyne Toussaint, disappeared after being attacked by demonstrators. On 13 November the antenna of Catholic radio station Voix de L'Ave Maria was vandalised by demonstrators. (RSF)

Yvon Chery, director of Radio Télé-diffusion Cayenne (RTC), had rocks thrown at his car near the city of Les Cayes on 22 November. Chery had reported several threats made against him in November. (RSF)

HONDURAS

On 24 October police and soldiers opened fire on an unarmed group of 30 peasants, who had approached them requesting permission to recover crops planted on land from which they had been expelled. Three people were killed and five wounded. (*Mesoamerica*)

On 25 October Ramon Custodio, president of the Honduran Human Rights Committee (CODEH), denounced the murder of six alleged members of military intelligence during October. 'The extrajudicial executions are occurring at a time when their testimony is fundamental to clearing up the truth,' he said. Julio Fonseca, a former military intelligence officer, asked for protection from the Committee, saying that he feared he would be killed because of information he has regarding disappearances in the 1980s. Meanwhile, the three military officers wanted on charges of torturing students (*Index* 6/1995) have gone into hiding. The judge in the case, Roy Edmundo Molinas, has been receiving death threats. (SWB, Reuter, AI)

HONG KONG

In late October China's Preliminary Working Committee in Hong Kong caused uproar with its proposal to scrap the territory's Bill of Rights. It also proposed the reintroduction of harsh colonial ordinances rendered unlawful by the Bill, including restrictions on public meetings and broadcasting, bans on groups that have not been formally registered, and the use of emergency powers. Two weeks later the PWC recommended that émigrés returning to Hong Kong be stripped of the right to vote and barred from taking an active part in elections. (Reuter, *Economist*, *South China Morning Post*, *Guardian*)

On 30 October prisoner Chim Shing-chung successfully challenged prison authorities' right to remove racing supplements from newspapers. A high court judge described Stanley Prison's attempt to limit illegal gambling as interference in the freedom of the press and a violation of prisoners' rights. (Reuter, *South China Morning Post*)

On 14 November the BBC announced plans to close its Hong Kong transmitter, to prevent its use by Chinese authorities after 1997. The transmitter broadcast news in Mandarin of the Tiananmen Square massacre into China in 1989. It will be replaced by a new transmitter in Thailand. (*Guardian*)

A delegation of Hong Kong educationalists who visited Beijing in mid-November learned of forthcoming revisions to history and geography textbooks. Accounts of the Opium War will be subject to revision, more emphasis will be placed on Hong Kong's past within China, and detailed discussion of the Tiananmen Square massacre will be removed. (*Guardian*)

Hong Kong journalists were exhorted to 'love their mother country... at this historic moment' at a press symposium on 27 November. Zhang Junsheng, vice-director of Xinhua (New China News Agency), said that protection of press freedom was clearly defined in China's Basic Law and, referring to 'the self-censorship issue', suggested that 'a certain degree of discipline is necessary. Anybody who objects to this discipline is being unreasonable.' (*South China Morning Post*)

INDIA

The third Insat-2 satellite was launched on 6 December. The new satellite will reach the whole subcontinent and also eastern Europe, southeast Asia and Africa. Doordarshan, the state broadcasting company, controls the use of the satellite. (*Financial Times*)

The government refused on 14 November to grant an entry visa to the threatened Bangladeshi writer Taslima Nasrin. The official reason was concern for her safety, but observers believe the ruling Congress party fears alienating Muslim voters. (*Times*)

On 15 November the Supreme Court ordered the Central Bureau of Investigation (CBI) to look into the disappearance of Jaswant Singh Khalra, a human rights activist. Police deny that Khalra was taken into custody on 6 September. This is the latest of a series of disappearances in Punjab state (*Index* 1/1995). (AI)

Recent publications: *Determining the Fate of the Disappeared in Punjab* (AI, October 1995, 16pp); *Torture Continues in Jammu and Kashmir* (AI, November 1995, 11pp)

INDONESIA

In a decision dated 11 October, but not released until 24 November, the High Court increased the sentences of Ahmad Taufik and Eko Maryadi (*Index* 4/1995, 5/1995, 6/1995). The two had sought to have their 32-month sentences overturned on appeal. The sentences were increased by four months each, on the basis that their actions conflicted with 'the role of the press, which is to enhance national security and to increase national awareness'. The journalists were denied legal representation. The court also announced it was upholding the 20-month sentence imposed on Danang Kukuh Wardoyo (*Index* 4/1995, 5/1995, 6/1995), and the two-year sentence given to Tri Agus Susanto, editor of *Kabar Dari Pijar* (News from Pijar) (*Index* 3/1995, 6/1995). (CPJ)

The newspaper *Suara Karya* reported on 1 November that the Java Military Command had arrested 11 members of the Islamic State of Indonesia (NII) and seized 95 books on 'deviationist teachings'. The books contained comments 'contrary to true Islamic teaching'. (SWB)

On 3 November the film censor announced that Chinese songs and characters are allowed to be broadcast on television. The use of Chinese in teaching and printed material was banned after the 1965 coup attempt, blamed on the Communist Party. (Reuter)

On 9 November eight dignitaries were prevented from entering East Timor by Indonesian officials. The delegation included Bishop Soma from Japan and two Irish members of the European Parliament. (Reuter, *Irish Times*)

On 22 November the Administrative Appeal Court upheld a decision taken in May by the Jakarta Administrative Court, which found that information minister Harmoko should rescind the ban on the weekly magazine *Tempo*. The court reaffirmed that the 1984 ministerial decree under which Harmoko banned the magazine in June 1994 conflicts with 'higher laws', a reference to the Press Act of 1982 and its guarantee of protection against censorship and closure. (*Straits Times*, SWB)

Information minister Harmoko has suggested that the government's main concern with the Internet is politics rather than pornography. In a statement to Parliament on 6 December he indicated that he would monitor the Internet for matters 'harmful to national security'. (*South China Morning Post*)

Recent publication: *Irian Jaya — National Commission on Human Rights Confirms Violations* (AI, September 1995, 13pp)

IRAN

On 28 October the editor of the daily *Tous*, Mohammed Sadeq Javadi Hessar (*Index* 6/1995), was convicted of slander and divulging secrets. Hessar's trial was conducted without a jury, and he was sentenced to six months in prison and 20 lashes. (CPJ)

Writer Sayed Morteza Shirazi was arrested on 21 November during a round-up of followers of his father, the Grand Ayatollah Shirazi who fled Iraq in the 1970s. Sayed Morteza Shirazi has written books chiefly on economics. His current whereabouts are unknown. (PEN)

ISRAEL

Mordechai Vanunu (*Index* 6/1994, 3/1995, 4/1995), now in his ninth year of solitary confinement in Ashkelon jail near Tel Aviv, is believed to be suffering from mental illness, according to his family. He has been offered improvements in jail conditions if he agreed not to discuss his kidnapping from Italy by Israeli secret services in 1986, but has refused. (*Guardian*)

JAPAN

On 30 October the Tokyo District Court issued an order to disband the Aum Shinrikyo cult (*Index* 6/1995) under the Religious Corporation Law. (Jiji Press Newswire)

KENYA

Legislation is rumoured to be pending for the creation of a Press Council and the establishment of a government Mass Media Commission. The proposed legislation would apparently require that all journalists, broadcasters and publications in Kenya obtain government licences. A set of punitive measures for 'professional misconduct' would also be introduced. The ruling KANU party is also expected to increase its efforts to control the media as the multiparty elections in 1997 draw closer. (International Press Institute)

President Moi dismissed Ugandan claims that Kenya had amassed troops at the border with that country as 'ridiculous and untrue' on 14 November. He went on to criticise a section of the local press for fuelling the claims

and urged Kenyan journalists to behave responsibly. He said that it was unfortunate that a number of local journalists were 'unpatriotic' and bent on 'jeopardising the interests of the motherland'. (SWB)

In a report on Kenyan TV on 22 November the foreign minister, Kalonzo Musyoko, accused foreign and local media representatives of portraying Kenya in a negative light. He said that Kenya was entitled to better treatment which would only come about through the dissemination of factual information. He specifically mentioned articles in *The Economist* and the *Nairobi Law Monthly*, which he accused of practising 'cheap journalism'. (SWB)

KYRGYZSTAN

On 2 October the Ministry of Education and Science issued an order banning both religious and atheist instruction in all state schools and universities. (SWB)

Six presidential candidates signed a statement on 8 November, accusing the media of bias in favour of President Akayev in the run-up to the elections of 24 December. They said that, in effect, the media are promoting just one candidate, Akayev. (SWB)

LEBANON

Aouni al-Kaaki, owner of the daily *As-sharq*, was sentenced in absentia to one month in jail after his newspaper published three cartoons that were deemed offensive to President Elias Hrawi and his wife Mona in 1994. (Reuter)

LITHUANIA

A bomb exploded at the office of Lithuania's largest newspaper, *Lietuvos Rytas*, on 16 November, causing considerable damage. The newspaper had been reporting on organised crime and it is widely believed that these reports provoked the attack. (IFJ)

MALAWI

The publisher and editor of the newspaper *Tribute*, Akwete Sande, was arrested on 22 November after refusing to reveal his source for an article in which he alleged that Vice-President Justin Malewezi's bodyguards were plotting to assassinate the president. (MISA)

MAURITANIA

At least 50 people, including Mohammed Ould Hamady, former director-general of national television, Abdellahi Ould Mohamedou, director-general of national radio, and Mohammed Ould Bowba, editor-in-chief of the journal *Akhbar El Ousbou*, were arrested around 24 October on suspicion of providing information to Iraqi secret services. Many of those arrested are connected to Iraq's ruling Ba'th party and a Ba'th splinter party Attali'à. They are being held incommunicado. (AI)

Three editions of the weekly *Le Calame* have been seized by the Ministry of the Interior: the French-language edition was seized on 12 November and the proofs of the Arab-language editions were confiscated at the printers on 22 and 29 November. No reason has been given. (RSF)

MEXICO

On 4 October three members of a vaccination team were gang-raped near San Andrés Larraínzar, Chiapas. A local women's advocacy group, Women of San Cristóbal, has recorded 50 rapes in the last 18 months. Cecilia Rodríguez, US representative for the rebel Zapatistas, was raped and sodomised by three armed men on 26 October in Chiapas. According to Rodríguez, sexual assault is being carried out as part of the government's low-intensity warfare strategy against the civilian population in the region. (*Latinamerica Press*)

On 15 November Amnesty International published a report on human rights violations, documenting 40 cases of extrajudicial execution by the security forces and 35 cases of torture so far this year. (AI)

Recent publication: *Human Rights Violations in Mexico — A Challenge for the Nineties* (AI, November 1995, 54pp)

MOROCCO

The French weekly magazine *Jeune Afrique* has been banned until further notice following the seizure of the 23 November edition which carried an article on Morocco headlined 'Uncertain Times'. (RSF, Reuter)

Kawtar Bahi, the daughter of journalist Mohammed Bahi who has been held by the Western Saharan liberation organisation Polisario for 10 years, called for his release in a press conference on 21 November. Mohammed Bahi, who had worked for the government paper *al-Anbaa*, was arrested by Algerian forces while covering the war in Western Sahara in 1986 and handed over to Polisario. It is believed that he is being held in a desert camp and that his health is deteriorating. Polisario released 186 Moroccan prisoners of war on 17 November on humanitarian grounds. (SWB, Reuter)

MOZAMBIQUE

On 16 October President Joaquim Chissano issued a decree creating a new Information Office, which will be part of the prime minister's office. The National Information Directorate is abolished, and its powers transferred to the new office. (SWB)

The Sunday newspaper *Domingo* and the chief of police, Domingos Maita, have been charged by the publisher of the weekly *Savana* with defamation, following reports in *Domingo* written by Maita, that *Savana* journalists were receiving payments and favours from an alleged drug trafficker. (MISA)

NAMIBIA

The Namibian Broadcasting Corporation (NBC) suspended a live call-in radio programme on 16 November, after allegations that death threats were being issued by callers to government officials in connection with the government's decision not to reward former combatants in the fight against South African occupation with cash payments. (MISA)

The prosecutor-general, Hans Heyman, may sue labour minister Moses Garoeb over remarks he made at a press briefing in November. The minister is alleged to have accused labour courts of having become 'rubber stamps' for employers. (MISA)

NIGERIA

Ken Saro-Wiwa (*Index* 2/1995, 5/1995, 6/1995) and eight other Ogoni rights activists — Barinem Kiobel, Saturday Dobue, Paul Levura, Monday Eawo, Felix Nwate, Daniel Kbakoo, John Kpunien and Baribo Bero — were hanged at Port Harcourt prison on 10 November. The nine were sentenced to death on 31 October after being convicted of inciting the murder of four Ogoni leaders, despite little evidence having been presented to support the charges. Two prosecution witnesses stated that they had been bribed to testify against Saro-Wiwa. (PEN, CCPJ, Reuter)

NORTH KOREA

On 24 November the president of the Workers' Party of Korea published a long article in *Nodung Simmun*, stating that the basic duty of the press is to propagandise in favour of the 'leader's greatness'. (SWB)

PAKISTAN

The Karachi offices of the *News International* were fired on by unidentified gunmen with automatic weapons on 22 October. Windows were shattered by bullets but no-one was hurt. The Newspapers' Editors' Council of Pakistan (NECP) reported that security forces were nearby during the shooting, but made no effort to intervene. (Pakistan Press Foundation)

Police raided the Hyderabad offices of the *News* on 4 November. Ansar Naqvi, a correspondent, was threatened with arrest and the office was ransacked. Police claimed they were searching for members of the Mohajir Quami Movement (MQM). (Pakistan Press Foundation)

Mujibir Rehman, a photographer for the *Star* and president of the Pakistan Association of Press Photographers, was beaten by police officers at a Karachi police station on 19 November after covering a demonstration by women against the detention of children. (Pakistan Press Foundation)

The bullet-riddled bodies of Nasir Hussain and his son Arif, the brother and nephew respectively of MQM leader Altaf Hussain, were found in Karachi on 9 December, several days after the men had apparently been taken into custody by the Paramilitary Rangers. The MQM blames the security forces for the murders, a charge they deny. The authorities have ordered an enquiry into the killings. (Reuter)

By the end of 1995 more than 1,730 people, including 190 policemen, had been killed since the beginning of the year in Karachi's ongoing civil violence. Eight hundred people were killed in 1994. (Reuter)

PALESTINE (GAZA-JERICHO)

Palestinian media were 'cautioned' by the Palestinian National Authority (PNA) not to publish any expressions of joy at the assassination of Israeli prime minister Yitzhak Rabin on 4 November. PNA spokesman Marwan Kanafani said he had urged the press 'to be more responsible in their reporting'. (Reuter)

The editors of two Islamist opposition newspapers, Alaa al-Saftawi of *al-Istiqlal* and Ghazi Hamad of *al-Watan*, were arrested by Palestinian police on 27 November in Gaza. Police chief Ghazi al-Jabali said that al-Saftawi had been arrested for incitement but gave no reason for Hamad's arrest. (Reuter, Palestinian Media Monitoring Centre)

Iyad Saraj, the human rights commissioner of the Palestinian Independent Commission for Citizen's Rights, was detained by PNA police in Gaza City on 7 December. It is unclear whether any charges have been brought against him.

The day before his detention, Saraj had attended a conference where he voiced severe criticism of the PNA's human rights record. (Palestinian Centre for Human Rights)

PANAMA

In November the non-governmental group Services of Peace and Justice (SERPAJ) released photographs of Guaymi men, women and children being restrained in pillories for 'breaking the law'. SERPAJ co-ordinator José Mendoza said that such punishment is usually meted out to those involved in disputes with non-indigenous ranchers over Guaymi land in Chiriquí Province. SERPAJ plan to take the case to the United Nations. (*Mesoamerica*)

PARAGUAY

César Barrios, a member of the Conscientious Objection Movement (MOC), was abducted and tortured on 4 November while travelling to the town of Pirapey, department of Itapua. The bus Barrios was travelling on was stopped by five soldiers. He was singled out among the passengers and asked for his military exemption papers, which the soldiers then ripped up before taking him to military headquarters in Ciudad del Este. Barrios was tortured and questioned about the MOC until the early hours of 5 November, when a soldier helped him to escape. (AI)

PERU

The government is to pay US$1.33 million to relatives of nine students and a professor who were kidnapped and executed by an army death squad in July 1992, it was announced on 27 October. The payment for the La Cantuta University slayings is the government's first acceptance of responsibility for extrajudicial executions in 15 years of guerrilla war. The officers accused in the massacre were jailed by a military court early in 1995 but freed under an amnesty introduced for military personnel. (Reuter)

Journalist Jesus Alfonso Castiglione Mendoza was sentenced to 20 years' imprisonment for membership of the Shining Path terrorist group on 21 November. The defence immediately filed an appeal for the sentence to be revoked. Castiglione has been detained since 29 April 1993 and charged on the basis of circumstantial evidence. His sentence was originally annulled in June 1995, after which the retrial was ordered. (Instituto Prensa y Sociedad)

Armed assailants hijacked a truckload of newsprint bound for the daily *La República* on 4 December. The men, who identified themselves as members of the government agency responsible for policing taxation, said the truck's cargo was being seized because it did not have a waybill, even though no waybill is required for the purchase of newsprint. (Instituto Prensa y Sociedad)

Recent publications: *The Peru Reader — History, Culture, Politics* (Latin American

Bureau, November 1995); *Women in Peru — Rights in Jeopardy* (AI, November 1995, 21pp)

PHILIPPINES

Two hundred and sixteen political prisoners are reported to remain in jails across the Philippines. Human rights group Task Force Detainees of the Philippines also record 14 cases of torture and seven summary executions carried out since September 1994. (*Philippine Human Rights Update*)

Doubts have been expressed about draft anti-terrorism legislation proposed by President Ramos. It has been suggested that the law, first discussed in July, could be used to criminalise political dissent, prohibiting peaceful protest and the possession of publications critical of the government. A draft revision of the Constitution by General José Almonte could also pose a similar threat: human rights safeguards have been omitted, habeas corpus can be suspended 'when the public safety requires it', and 'enemies of the state' will be subject to longer periods of detention. (*Philippine Human Rights Update*)

POLAND

On 15 October then-President Walesa accused Adam Michnik, editor-in-chief of the daily *Gazeta Wyborcza*, of 'highly unethical manipulation' in publishing the 'unauthorised text' of an interview with him, which, Walesa, says was in 'too frank and colloquial' a form'. (SWB)

The Polish Episcopate Commission for the Media warned the Catholic media on 23 October against defaming candidates standing in the presidential elections. It said that some media — particularly Radio Maryja — have been propagating 'anti-Semitic stereotypes' and reproaching candidates for their Communist past. (SWB)

ROMANIA

On 24 October the Chamber of Deputies deleted references to the media from proposed revisions of the Penal Code relating to defamation (*Index* 6/1995), so that journalists will no longer be placed in a separate category. The new amendments impose prison sentences of up to two years for libel and three years for disseminating false information. (OMRI)

It was reported in November that physicist Silviu Axinescu has been denied permission to travel abroad for professional purposes by his employer, the Institute of Atomic Physics in Bucharest. He has also reportedly suffered constant harassment by the Institute's trade union, which is said to be under government control, apparently for maintaining contact with foreign scientists without the agreement of union executives. His employment has allegedly been threatened, his mail opened and his fax and e-mail facilities controlled or cancelled. He has also been accused of damaging Romania's image by criticising the way scientific research is funded. (AAASHRAN)

In November Hungarian media accused the Romanian National Audio-Visual Council of banning cable network transmission by Duna TV, a Hungarian satellite station, by refusing to issue licences to any cable companies. The Council denies that the decision was aimed specifically at Duna. (OMRI)

Recent publication: *Romanian Authorities Respond to Amnesty International's May 1995 Report* (AI, October 1995, 10pp)

RUSSIAN FEDERATION

Russia: Turkmen poet Shiraly Nurmuradov (*Index* 6/1995), winner of the Tchkolsky award for writers in exile, has taken up his year's residency in Sweden, offered as part of the prize. The criminal case against him in Russia, however, remains open. (PEN)

The editor-in-chief of *Literaturnyi Vesti* newspaper was beaten up by unknown assailants at his home on 5 November, and papers pertaining to preparations for an anti-fascist congress were stolen. He was taken to hospital with 'serious injuries and concussion'. (SWB)

On 5 December a small bomb destroyed the office of Nikolai Lysenko, the leader of the extreme right-wing National Republican Party and a deputy in the Duma (the lower house of Parliament), fuelling fears of further political violence in the run-up to

INDEX INDEX

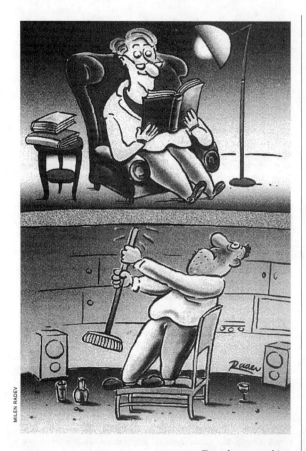

Everyday censorship

the 17 December parliamentary elections. (*Times*)

Chechnya: A bomb exploded at the headquarters of the Moscow-installed government in Grozny on 4 December, killing 11 people. The explosion occurred hours after General Dzhokar Dudayev, the Chechen separatist leader, pledged his commitment to finding non-violent solutions to his country's problems. (*Times*)

RWANDA

Jean Rubaduka, a member of the Council of State and a regular contributor to the Catholic bi-monthly *Kinyamateka*, survived a second attempt on his life in the early hours of 28 November. An intruder broke into his house, but was foiled by the intervention of Rubaduka's neighbours. Three days earlier four assailants had attacked his home at about 3.30am and again only fled when a neighbour answered his cries for help. (RSF)

There are fears for the safety of André Sibomana, editor-in-chief of *Kinyamateka*, after a military intelligence chief urged the authorities to arrest a number of Catholic priests — including Sibomana — at a Department of Military Intelligence meeting in mid-November. (RSF)

SAO TOME & PRINCIPE

On 26 October Manuel Barros, editor-in-chief of the daily *Noticias*, was placed under house arrest, awaiting trial reportedly in connection with a recent article entitled 'VIP Treatment for Drug Traffickers', in which he criticised judges' attitude towards the drug trade. Although the house arrest order was lifted on 27 October, Barros must still notify the authorities of his movements. (RSF)

Oscar Medeiros, a journalist with state television, was arrested at his home on 21 November. The charges against him were not made public, but the arrest occurred three days after Medeiros had received a statement from Antonio Quintas, secretary of state for information and culture, which was supposed to be read in full on a news bulletin. Medeiros read only two paragraphs. On 22 November, Medeiros was released from detention, and on 23 November the complaint by Quintas was withdrawn. (RSF)

SERBIA-MONTENEGRO

Kosovo: Serbian police reportedly raided ethnic-Albanian 'shadow-state' schools and universities in Urosevac and surrounding villages in October. (SWB)

On 16 November Astrit Saliu, a journalist with *Koha,* was reportedly arrested at the paper's offices and detained and beaten at a police station before being handed over to the state security police. His detention was related to evidence he had gathered which resulted in the withdrawal of terrorism charges against Ganimet Podvorica on 10 November. Saliu was interrogated by security officers over his paper's stance on political developments in Kosovo. (SWB)

On 27 November Serbian police in Pristina dispersed thousands of Albanian demonstrators attending 'Day of the Flag' independence celebrations. Over 50 people were arrested, maltreated or imprisoned by police. (SWB)

Montenegro: On 10 November it was reported that *Monitor* journalist Vladimir Jovanovic was arrested and prevented from leaving the country to attend a media workshop in Slovenia, apparently for possessing false documents. (OMRI)

Serbia: On 11 October *Nasa Borba* journalists ended a week-long strike after the paper's management gave journalists the right to endorse the appointment of a new editor. The strike was held in protest at a unilateral management decision to amend editorial statutes, which had previously guaranteed journalists the right to elect the paper's editor. (IFJ)

On 9 November *Nasa Borba* reported that a satirical weekly, *Smrklost,* is to be launched in Kragujevac. It will be the first of its kind in Serbia and has been likened to Croatia's *Feral Tribune,* which lampoons political and social developments. (OMRI)

SINGAPORE

Attorney-general Chan Sek Keong has criticised lawyers who fail to defend the country's legal system against attacks by the foreign media. In a speech to the Law Society, Chan said lawyers who question the government's tough stand on crime 'can only reinforce the absurd thinking of libertarian academics and the Western liberal press that no accused can get a fair trial in Singapore.' (*Far Eastern Economic Review*)

The *International Herald Tribune* has agreed to pay US$200,000 to senior minister Lee Kwan Yew over an article by US academic Christopher Lingle, published on 7 October 1994. The article referred to unnamed 'intolerant regimes in the region', some of which rely 'upon a compliant judiciary to bankrupt opposition politicians'. (Reuter)

On 29 November five Jehovah's Witnesses were fined between US$1,400 and $2,800 for attending a meeting of a banned organisation. Two of the five refused to pay and were given two-week jail sentences. The case stemmed from a raid in February 1995, when 65 Jehovah's Witnesses were arrested and literature seized. The movement was proscribed in 1972 for being 'prejudicial to public welfare and order'. (Reuter)

SLOVAKIA

On 23 October Bratislava Municipal Court found prime minister Vladimir Meciar responsible for slandering former dissident Vladimir Pavlik at a rally in May 1990. Pavlik, a signatory of Charter 77 and an editor of the radical anti-Communist *Necenzurovane noviny* (Uncensored News), had attempted to reveal widespread theft from an arms factory. Meciar, then interior minister of Slovakia (which was still part of the federal Czechoslovak state), had described Pavlik's efforts as 'seditious' and cast aspersions on his intelligence. The court ruled that Meciar must give a written apology and pay him US$690. (SWB)

A new bill on the audio-visual industry, approved by the government on 24 October, requires, inter alia, that Czech films, or foreign films dubbed into Czech, must have Slovak subtitles or be dubbed into Slovak. On the same day, the government also approved a new version of the state language law, which culture minister Ivan Hude said will restore the rights of Slovaks and correct a situation in which the Hungarian ethnic minority had 'too many rights'. Leaders of the 500,000-strong Hungarian minority warned that the law will not only jeopardise the rights of that minority but may also affect the opposition press. (SWB)

SOUTH AFRICA

The South African Broadcasting Corporation (SABC) rejected an advertisement scheduled to be aired on Radio Highveld, for the weekly *Mail & Guardian*. The advertisement was banned on grounds that it was blasphemous, although it had previously been broadcast without complaint on Radio 702. (Freedom of Expression Institute)

The SABC withdrew an episode of the Afrikaans drama *Die Laksman* on 12 October, after the Atomic Energy Corporation (AEC) gained an injunction prohibiting transmission. The AEC's main concern was the similarity between names of characters in the series and real people. The drama centres on spying and the illegal trade in atomic weapons. (*Natal Witness*, Freedom of Expression Institute)

Gauteng premier Tokyo Sexwale mounted a vigorous attack on the media on 19 October, Press Freedom Day. He described the two best-selling Sunday newspapers, the *Sunday Times* and *Rapport*, as racist and the *Mail & Guardian* as 'a pseudo-liberal English weekly which masquerades as a progressive paper'. (*Southern Africa Report*, *Mail & Guardian*)

SOUTH KOREA

Journalist Choi Chin-sop was released on 15 November after just over three years in prison. Choi, who worked for the pro-democracy magazine *Mal*, was charged under National Security Law with being part of a North Korean spy ring. (RSF)

On 1 December Pak Yong-kil (*Index* 5/1995, 6/1995) was found guilty of violating National Security Law and sentenced to a year in prison, suspended for three years owing to ill health. Her civil rights are suspended for two years. (AI, SWB)

Recent publication: *International Standards, Law and Practice — The Need for Human Rights Reform* (AI, November 1995, 62pp)

SRI LANKA

On 7 November the Supreme Court rejected a petition from Wimal Wickramasinghe, a former United National Party cabinet member and publisher of *Jana Jaya*, to lift the government's ban on reporting military news (*Index* 6/1995). (Reuter)

In the government attack on rebel LTTE positions around Jaffna, 320 government troops and 1,500 rebels had been killed up to the beginning of December, according to government figures. Some 300,000 civilian refugees have fled the war zone. The government continues to restrict media coverage (*Index* 6/1995): foreign news teams were not permitted to approach the front lines and local newspapers had sections of articles deleted by the censor. (*Independent*)

SUDAN

In the wake of recent demonstrations (*Index* 6/1995), the National Islamic Front's security forces arrested two journalists, Mohammed Abd al-Khaliq and Hudan Mahjoub, in late October. Instructions have also been issued to arrest 11 others who are jointly accused of masterminding the riots and publishing *al-Shabiba*, the mouthpiece of the Sudan Youth Union. (*Sudan Update*)

SWAZILAND

An article about the 27-year reign of King Mswati, entitled 'The Playboy King', in the August 1995 issue of *Playboy* led to a renewal of the 1973 proscription order on pornographic publications. Police subsequently carried out raids on shops that stock pornographic publications. (MISA)

Two senators, Mark Ward and Walter Bennett, are calling for a government bill to prevent the press from criticising the king and to establish a press council. The demand came after criticisms were published in the Swaziland *Times* about the size of the delegation which accompanied King Mswati to the USA for the 50th anniversary of the UN. (MISA)

SYRIA

The health of poet and journalist Faraj Ahmed Beraqdar

(*Index* 3/1993, 1 & 2/1994) is believed to be deteriorating as a result of torture and the lack of medical treatment in prison. Arrested in 1987 on suspicion of membership of the banned Party for Communist Action (PCA), he was held incommunicado for seven years before being tried and sentenced to 15 years' imprisonment in 1993. Former army officer Khalil Brayez (*Index* 6/1994) remains in jail, despite his sentence having expired in 1987. He was arrested in Beirut in November 1970 and sentenced to 15 years' imprisonment in March 1972 on charges of 'revealing state secrets' in two books that he wrote about the Syrian army's activities in the Golan Heights in the 1967 war. (PEN)

TANZANIA

The first multi-party elections held on 29 October throughout Tanzania's mainland and Zanzibar islands degenerated into chaos, with voting in the capital, Dar Es Salaam, postponed until 19 November. Elsewhere on the mainland the process collapsed almost entirely as the electoral commission failed to get ballot papers, officials or voting papers to booths on time. The worst problems occurred in opposition strongholds, fuelling suspicions that the chaos was orchestrated. The ruling Party for the Revolution (CCM) eventually won and Benjamin Mkapa was elected president. (*Economist, Financial Times, Observer*, Reuter)

At a ceremony to launch a satellite station in Dar Es Salaam on 4 November then-President Ali Hassan Mwinyi called on media institutions to provide proper training for their journalists. He stressed the need for journalists to file correct and true reports in order to avoid creating 'unnecessary conflicts in society'. (SWB)

THAILAND

On 23 November the defence minister, Chavalit Yongchaiyudh, ordered military leaders to withdraw radio talkback programmes from stations under their control if they are used to criticise the government. The programme *Joh Khao Haa Thum* was banned on 18 October for featuring political criticism which 'tarnishes the country's image'. The show's host, Suwat Yomchinda, has said he will take legal action against the government over the ban. (*Bangkok Post*)

On 29 November U Ye Gyaung and his wife, Daw Khin Hlaing, were arrested in Bangkok. U Ye Gyaung is a former editor of several magazines and newspapers in Rangoon. He currently helps to edit the *New Era Journal*, a bi-monthly opposition Burmese newspaper published in the 'liberated area' on the Thai-Burmese border. They were among 25 Burmese asylum-seekers arrested in a crackdown on Burmese dissidents in Thailand. (AI, *Bangkok Post*)

TIBET

Chadrel Rinpoche (*Index* 5/1995) was formally stripped of his position as abbot of Tashilunpo monastery in Lhasa, seat of the Panchen Lama, in September. Other members of the monastery were also removed from their posts. (*South China Morning Post*, Reuter)

On 5 November China summoned 75 senior Tibetan lamas to Beijing to begin the search for a second Panchen Lama to replace the Dalai Lama's disputed choice (denounced by China as null and void). Tibet's government-in-exile has accused China of coercion and many of the 75 lamas urged Beijing to reinstate the Dalai Lama's choice, Gedhun Choekyi Nyima. On 29 November Gyaltsen Norbu was named as the new Panchen Lama in Beijing. China denies that Gedhun is still being held under house arrest in the capital. State television quoted Lhasa residents as praising Beijing's choice. (*South China Morning Post*, Reuter, SWB)

TOGO

On 21 November Fulbert Altisso, editor-in-chief of *L'Eveil du Peuple*, was arrested, without a warrant, for publishing an article on the killing of four civilians by the military on 29 September. If convicted, he faces a prison term of up to three years and could also be fined. (RSF)

TUNISIA

Several editions of 12 foreign publications carrying news articles about Tunisia have been banned from distribution since July. Italian dailies *La Stampa* and *Il Corriere della Sera* were banned in July and August after they carried articles about corruption in Tunisia. During French president Jacques Chirac's visit in October, several French publications were banned, including *Le Monde*, *Libération*, *Le Figaro*, *Jeune Afrique*, and *Courrier International*. In September the authorities refused permission for Reporters sans Frontières to investigate press freedom in Tunisia on the grounds that such an investigation was unnecessary. (RSF)

Mohammed Mouadda, leader of the opposition party Movement of Democratic Socialists (MDS), was arrested on 9 October and charged under the Penal Code with treason and conspiring with a foreign power against the state. Two of the charges carry the death penalty. His arrest came the day after MDS published a letter that they had sent to the president some weeks before, raising concerns about the current political situation and restrictions on political parties' activities. (AI)

Mohammed Kilani (*Index* 2/1995, 4/1995), a leading member of the Tunisian Workers' Communist Party (PCOT) and former editor of PCOT's newspaper *El Badil*, and Hamma Hammami (*Index* 4/1992, 6/1992), a director of *El Badil*, were released in early November, after receiving a presidential pardon on the anniversary of President Ben Ali's accession to power. Mohammed Hedi Sassi, sentenced to four years' imprisonment in 1994 for distributing PCOT leaflets (*Index* 3/1994), and Mohammed Najib Hosni, a human rights lawyer who was arrested in June 1994 (*Index* 4/1994), remain in jail. (AI)

Recent publication: *Repression Thrives on Impunity* (AI, November 1995, 40pp)

TURKEY

Haluk Gerger (*Index* 1/1995, 5/1995, 6/1995) was released from prison on 27 October. Gerger completed a 20-month prison sentence in September, but was not released because he had refused to pay a large fine, which was converted into a three-year sentence. He has since paid the fine and been released. (AAASHRAN)

On 29 October 14 employees from the weekly *Kurtulus* (Liberation) were arrested. Bulent Bagci, Nuray Yucel, Neslihan Uslu, Nilufer Gunes, Halil Aksu, Muhittin Erdogan, Fikri Hidirlar, Ilker Alcan, Yasar Aslan, Mahir Dagdeviren, Iluseyin Kilic, Tarik Tolunay, Alaadin Koyun and one unnamed person were taken to Aksaray Anti-Terror headquarters. Police seized 8,000 copies of that week's edition of *Kurtulus* (no16), which they were about to deliver to the news-stands. The remaining copies of the issue were seized at the printing house. All detainees were released on 7 November. Yucel and Tolunay were allegedly tortured while in detention. (RSF)

During the early weeks of November Turkish courts ordered the release of 123 people after Parliament passed an amendment to Article 8 of the Anti-Terror Law. The amendment allows courts to free people jailed for separatist propaganda before their terms are completed. There is no indication how many of the people ordered to be released have actually left prison. (Reuter)

On 3 November Necmiye Arslanoglu and Nuran Tekdag, journalists with *Ozgur Halk* (Free People), and Metin Acet, journalist with *Ozgur Politika* (Free Policy), were detained and taken to Batman police headquarters. Arslanglu, who has been detained on four previous occasions (*Index* 3/1995, 6/1995), has since been released. Metin Acet and Nuran Tekdag have been formally arrested. (AI)

On 9 November Aliza Marcus, Istanbul correspondent for Reuters news agency, was acquitted of charges of inciting racial hatred (*Index* 6/1995). She had been charged in connection with an article she wrote in November 1994, which described the forced evacuation of Kurdish villages in southeastern Turkey. (CPJ)

The Radio and TV High Council closed Interstar and Kanal-6 television stations for three days on 17 November, after Interstar failed to air the 'corrected version' of a reply by prime minister Tansu Ciller, her husband and others, to accusations of corruption. Kanal-6, meanwhile, had broadcast claims that some members of the government and Parliament are homosexuals. (SWB)

On 24 November Ismail Besikci, a sociologist who has already received sentences of around 200 years in prison for writing about the Kurds and Kurdistan (*Index* 1/1995), was sentenced to another six years in prison on six separate charges, despite the revision of the Anti-Terror Law. The Ankara state security court also sentenced Besikci's publisher, Unsal Ozturk, to a year in jail. Both had been charged under Article 8 of the Anti-Terror Law (Reuter)

Nazi Tavlas, editor of *Strategy News Bulletin*, is being prosecuted for 'revealing state secrets' in articles alleging corruption in the air force's procurement policy. Tavlas denies citing information from classified sources and maintains that his articles were based on public information. (CPJ)

Yasar Kemal was found not guilty of spreading separatist propaganda on 1 December. Kemal had been prosecuted for writing an article in *Der Spiegel* in which he severely criticised state policy towards the Kurds (*Index* 1/1995, 2/1995). (Reuter)

On 7 December the Istanbul State Security Court charged 99 writers, intellectuals and journalists for putting their names to an anthology of writings entitled *Freedom of Thought* in March 1995 (*Index* 4/1995). Most were charged under the Anti-Terror Law, but some were charged with offences such as 'inciting hatred' and 'apology for crime'. (RSF)

Recent publication: *Weapons, Transfers and Violations of the Laws of War in Turkey* (HRW, November 1995, 172pp)

TURKMENISTAN

It was reported in November that two journalists, Mukhamed Muradly, formerly with the journal *Diyar*, and Yovshan Annakurban, have been arrested apparently in connection with an anti-government demonstration which took place in July 1995. There are unconfirmed reports that Muradly has been tried and sentenced to a prison term for 'preparing and distributing printed matter opposing the president'; it is unclear whether Annakurban has also been tried. (AI)

UGANDA

Haruna Kanaabi (*Index* 5/1995), editor of the Islamic opposition newsletter *Shariat*, has been charged with an additional offence of publishing false news, which carries a 10-year sentence. Kanaabi has been in detention since 25 August on charges of sedition and was arrested with journalist Al-Haji Musa Hassein Njuki, who died in custody. Kanaabi is said to be living in poor conditions and is in ill health. Requests for bail have been denied. (CPJ)

UNITED KINGDOM

Islamic extremist groups were blamed by the National Union of Students on 30 October for a rise in racial tension on university campuses. In recent months over one hundred student unions have banned the Islamic group Hizb ut-Tahrir, also known as The Party of God. (*Times*)

The Sweden-based satellite service XXXTV - formerly known as TV Erotica - was effectively banned by the government on 14 November, when Virginia Bottomley, the national heritage secretary, announced an order outlawing the supply of smartcards necessary to unscramble its signal. (*Times, Independent*)

A whistleblower who helped the *Independent* newspaper expose a serious security breach at British Telecom was cleared on 14 November of illegally breaking into the company's main computer system. The decision in favour of Nigel Mahomet, a former BT engineer, effectively safeguards journalists' rights to receive secret computer information. (*Independent*)

Film censorship in Britain is to be exposed to court scrutiny for the first time, after Redemption Films was granted leave in November for a judicial review of procedures at the British Board of Film Classification (BBFC) and the Video Appeals Committee. The move follows a decision earlier this year to refuse classification to Redemption's *Bare Behind Bars*. Redemption argue that the BBFC was neither fair nor consistent in its approach to the video and that James Ferman, the BBFC's director, was manifestly wrong in his application

INDEX INDEX

of the new Criminal Justice Act's provisions relating to video classification. In the BBFC's 1994-95 report, issued in November, it was announced that six of the 3,500 videos submitted to it were banned while 6.5 per cent were ordered to have scenes cut before the films were granted certificates. (*Independent*)

On 27 November the government published its new draft charter for the BBC requiring stricter standards of taste and decency. Due to come into effect in the spring, the charter places greater pressure on the BBC's governors to monitor the Corporation's output, giving them a far more explicit regulatory role. The document also requires the BBC to draw up an impartiality code for the first time. (*Guardian, Observer, Independent*)

A video which includes controversial footage of a couple having sex in a lift was withdrawn from sale on 28 November, the second day of its release. *Caught In The Act*, a collection of closed-circuit TV footage compiled without the consent of those who appear on camera, was removed from the shelves by its producer, Barrie Goulding, after a storm of unfavourable press coverage about the threat it posed to civil liberties. It is believed that threats of legal action also forced the decision. (*Guardian, Independent*)

On 1 December the High Court ruled that the home secretary, Michael Howard, had acted unlawfully in his decision to ban the Reverend Sun Myung Moon from visiting Britain. Mr Justice Sedley made the ruling on the grounds that the founder of the Unification church had been denied the opportunity of stating his case before the exclusion order was made. (*Times*)

The December issue of a television listings magazine was withdrawn from the shelves of a number of stores because its problem page contained explicit advice to a sixteen-year-old girl on how to perform oral sex. *TV Hits* was withdrawn from sale at Asda, Sainsbury, Somerfield, Tesco and WH Smith after complaints from parents. (*Guardian, Independent*)

USA

The Senate has approved an amendment to the Telecommunications Act of 1995 (*Index* 4/1995, 5/1995) requiring television manufacturers to equip their sets with a blocking device known as a 'v' or 'c' chip. The chip would allow parents to block programmes based on encoded ratings similar to those used in the film and video industries. Manufacturers are however waiting until a standard is set by the broadcasters who are worried about the commercial impact of a ratings system on their higher-rated programming. (Interactive Home, Reuter)

In the wake of October's Million Man March, Cleveland District Court ruled on 2 November that Nation of Islam leader Louis Farrakhan can hold a men-only meeting at the city's convention centre against officials' wishes. The American Civil Liberties Union, representing the Nation of Islam, argued that sex discrimination laws were being used as pretext to deny a forum to a controversial speaker. (Reuter)

On 9 November lawyers acting for 14 Mississippi parents and their children said that a 1994 law allowing student-initiated non-sectarian prayers in public schools was unconstitutional and turned teachers into 'religious police'. The state judicial panel is expected to rule on the issue. (Reuter)

On 12 November CBS pulled its current affairs programme *60 Minutes* for fear of litigation from tobacco companies, even though no suit had been threatened. This 'self-censorship' follows ABC's decision in August to settle out of court in a $10 billion libel suit brought against them by the Phillip Morris tobacco company. Both ABC and CBS have recently been part of large media takeovers. In approving the takeover of CBS by the Westinghouse corporation on 22 November, one Federal Communications Commissioner expressed concern over the withdrawal of *60 Minutes* in an era of 'increasing media industry convergence'. (*Communications Today*, *Independent*, Reuter)

On 13 November the Supreme Court announced that it would rule on whether restrictions on indecent cable TV shows violate free-speech rights, with a decision expected by the end of the Court's term in June. The case stems from a ruling by the Appeal Court in June 1994 to uphold the 1992 Federal Communications Commission law requiring cable companies to keep indecent programming on a separate channel, unavailable until the subscriber requests it in writing. Lawyers acting for the cable operators described the law as an unconstitutional 'censorship scheme'. (Reuter)

On 4 December the Information Industry Association announced its support for a proposal submitted by Representative Rick White which would make it illegal for online service providers to knowingly make pornographic material available to minors. This follows the September FBI 'sting' operation which led to 12 arrests of America Online users for trading in child pornography via e-mail. However, AOL spokeswoman Pam McGraw argued against the prosecution of online services for pornography distribution, saying: 'That would be like saying the post office should be responsible for mailed child pornography.' The Department of Justice similarly argued that holding service providers liable would be 'fraught with problems', and would infringe their First Amendment rights. On 7 December a House-Senate conference committee provoked a storm of protest by approving a tougher-than-expected version of the telecommunications bill's 'online smut' provision, which would introduce hefty fines and prison terms for anyone who sends indecent words or images over computer networks without ensuring that they are not accessible to children. This latest version of the draft is closer to the original Senate proposal introduced by Senator James Exon (*Index* 4/1995). (PR Newswire, Netguide, *Los Angeles Times*)

Following the removal of advertisements for Calvin Klein jeans from the sides of city buses, New York City officials have forced Levi Strauss to remove its 'Nice Pants' advertising from the side of bus shelters. The advertisements featured a pair of jeans placed under plastic shields in 40 Manhattan bus shelters, underneath which was a message, visible if the trousers were stolen, which read: 'Apparently they were very nice pants.' A Levi Strauss spokesman denied the campaign was intended to encourage theft: 'The ads promote pants. Period. We're not

INDEX INDEX

in the business of promoting theft. We're in the business of promoting pants.' He added: 'The ad doesn't say "steal these pants". The ad says "Nice pants".' (*International Herald Tribune*)

Recent publications: *Banned Books Resources Guide* (American Library Association, 1995); *Children in Confinement in Louisiana* (HRW/Children's Rights Project, October 1995, 141pp); *Reintroduction of Chain Gangs* (AI, November 1995, 5pp)

VIETNAM

The Voice of Vietnam radio announced on 14 October that Hanoi customs officials had seized 253 'reactionary' magazines mailed from the USA, France and Canada. (SWB)

On 8 November it was reported that two veterans of the Vietnamese revolution have been jailed for writing articles calling for the rehabilitation of about 30 senior party officials purged during the 1960s for revisionism and opposing the party line. Hoang Minh Chinh was sentenced to 12 months and Do Trung Hiew was jailed for 15 months for 'anti-socialist pro-

paganda'. (SWB)

ZAIRE

Batabiha Bushoki, general secretary of the non-governmental Study and Action Group for Development (GEAD), is being held at a military camp in Goma, after taking questions from foreign journalists at a meeting of political parties and NGOs on 21 November. It is feared that he has been detained solely for expressing his opinions to the international media. (AI)

ZAMBIA

Two journalists from the *Sun* newspaper were dismissed on 12 October, following publication of an article criticising President Chiluba's administration. Joe Chilaizya and Jowie Mwinga have been warned in the past for 'unfavourable' reporting. (MISA)

ZIMBABWE

The resignation of Bill Saidi from the editorship of the *Sunday Gazette* in October has been linked to reprimands from the paper's publisher, Elias Rusike, over an article alleging President Mugabe had secretly married his former secretary. The article sparked off a criminal defamation charge against other newspapers carrying the same story (*Index* 5/1995). (MISA)

★★★

General publications: *The Human Rights Handbook — A Practical Guide to Monitoring Human Rights*, by Kathryn English and Adam Stapleton (University of Essex Human Rights Centre, 1995, 384pp. Price £15 from Dept AMM, PO Box 1995, Burgess Hill, W Sussex RH15 8QY, UK); *In the Public Eye — Parliamentary Transparency in Europe and North America*, ed Edwin Rekosh (International Human Rights Law Group, 1995, 297pp)

★★★

Compiled by: *Anna Feldman, Dagmar Schlüter, Kate Thal (Africa); Nathalie de Broglio, Dionne King, James Solomon (Americas); Erum Faruqi, Nicholas McAulay, Sarah Smith, Saul Venit (Asia); Laura Bruni, Robert Horvath, Robin Jones, Oleg Pamfilov, Vera Rich (eastern Europe and CIS); Michaela Becker, Philippa Nugent (Middle East); Jamie McLeish, Predrag Zivkovic (western Europe)*

★★★

Find Index at:

http://www.oneworld.org/index_oc/